Stereochemistry and Biological Acti[illegible]

Stereochemistry and Biological Activity of Drugs

Edited by

E.J. Ariëns *Director of the Institute of Pharmacology and Toxicology, University of Nijmegen, The Netherlands*

W. Soudijn *Division of Pharmaceutical Chemistry, Department of Pharmacy, University of Amsterdam, The Netherlands*

P.B.M.W.M. Timmermans *Division of Pharmacotherapy, Department of Pharmacy, University of Amsterdam, The Netherlands*

Blackwell Scientific Publications
Oxford London Edinburgh Boston Melbourne

Editorial offices:
Osney Mead, Oxford, CX2 OEL
8 John Street London, WC1N 2ES
9 Forrest Road, Edinburgh, EH1 2QH
52 Beacon Street, Boston
Massachusetts 02108, USA
99 Barry Street, Carlton
Victoria 3053, Australia

First published 1983

Set by Bakker - Baarn - The Netherlands

DISTRIBUTORS

USA
Blackwell Mosby Book Distributors
11830 Westline Industrial Drive
St. Louis, Missouri 63141

Canada
Blackwell Mosby Book Distributors
120 Melford Drive, Scarborough
Ontario, M1B 2X4

Australia
Blackwell Scientific Book Distributors
31 Advantage Road, Highett
Victoria 3190

British Library
Cataloguing in Publication Data

ISBN 0 632 01155 6

Contents

List of Contributors

E.J. Ariëns,
University of Nijmegen,
Department of Pharmacology,
and Toxicology
Geert Grooteplein N21,
6525 EZ NIJMEGEN/The Netherlands

A.J. Beld,
University of Nijmegen,
Faculty of Medicine,
Department of Pharmacology,
Geert Grooteplein N21,
6225 EZ NIJMEGEN/The Netherlands

M.G. Bogaert,
Heymans Institute of Pharmacology,
University of Gent Medical School,
De Pintelaan 185,
B-9000 GENT/Belgium

D.D. Breimer,
University of Leiden,
Department of Pharmacology,
Sylvius Laboratories,
Wassenaarseweg 72,
2333 AL LEIDEN/The Netherlands

A.S.V. Burgen,
National Institute M.R.,
Medical Research Council,
The Ridgeway, Mill Hill,
LONDON NW7 1AA/England

W.A. Buylaert,
Heymans Institute of Pharmacology,
University of Gent Medical School,
De Pintelaan 185,
B-9000 GENT/Belgium

R. Dahlbom,
University of Uppsala,
Department of Organic Pharmaceutical
Chemistry,
Biomedical Center,
Box 574,
S-751 23 UPPSALA/Sweden

C.A.M. Van Ginneken,
University of Nijmegen,
Faculty of Medicine,
Department of Pharmacology,
Geert Grooteplein N21,
6525 EZ NIJMEGEN/The Netherlands

G. Lambrecht,
University of Frankfurt,
Faculty of Biochemistry, Pharmacy and
Food Chemistry,
Department of Pharmacology,
Theodor-Stern-Kai 7,
Gebäude 75A,
D-6000 FRANKFURT/Federal Republic of
Germany

E. Mutschler,
University of Frankfurt,
Faculty of Biochemistry, Pharmacy
and Food Chemistry,
Department of Pharmacology,
Theodor-Stern-Kai 7,
Gebäude 75A,
D-6000 FRANKFURT/Federal Republic of
Germany

K.H. Rahn,
National University of Limburg,
Department of Internal Medicine and
Pharmacology,
Bio-medical Center,
P.O. Box 616,
6200 MD MAASTRICHT/The Netherlands

A.G. Rauws
National Institute of Public Health,
Unit for Pharmacokinetics,
P.O. Box 1,
3720 BA BILTHOVEN/The Netherlands

J.M. van Ree,
University of Utrecht,
Medical Faculty,
Rudolf Magnus Institute for
Pharmacology,
Vondellaan 6,
3521 GD UTRECHT/The Netherlands

J.F. Rodrigues de Miranda,
University of Nijmegen,
Faculty of Medicine,
Department of Pharmacology,
Geert Grooteplein N21,
6525 EZ NIJMEGEN/The Netherlands

R.R. Ruffolo,
Lilly Research Laboratories,
Department of Cardiovascular
Pharmacology,
307 East McCarty Street,
INDIANAPOLIS, Ind. 46285/USA

W. Soudijn,
University of Amsterdam,
Department of Pharmacy,
Division of Pharmaceutical Chemistry,
Plantage Muidergracht 24,
1018 TV AMSTERDAM/The Netherlands

P.B.M.W.M. Timmermans,
University of Amsterdam,
Department of Pharmacy,
Division of Pharmacotherapy,
Plantage Muidergracht 24,
1018 TV AMSTERDAM/The Netherlands

N.P.E. Vermeulen,
University of Leiden,
Department of Pharmacology,
Sylvius Laboratories,
Wassenaarseweg 72,
2333 AL LEIDEN/The Netherlands

A. Witter,
University of Utrecht,
Medical Faculty,
Rudolf Magnus Institute for
Pharmacology,
Vondellaan 6,
3521 GD UTRECHT/The Netherlands

Introduction

J.M. van Ree

The scope of this book is stereochemistry and biological activities of drugs, which was the topic of an international meeting held in Noordwijkerhout, The Netherlands, under the auspices of the Dutch Society of Pharmacology, the Medicinal Chemistry section of the Royal Dutch Chemical Society and the Dutch Society of Clinical Pharmacology and Biopharmacy.

The aim of that meeting was to start and stimulate a multidisciplinary discussion between scientists of different disciplines about the advantages and disadvantages, the possibilities and limitations in applying principles of stereoselectivity by preparing biological active molecules for therapeutic use in clinical practice. The meeting was opened with the message: 'Let us start to communicate and at the end of the meeting reflect where we stand'.

During the meeting it became evident that interest in stereoselectivity is not because we are dealing with the latest novelty in modern pharmacology, but because stereochemical principles can contribute to our understanding of the pharmacological action of drugs and to our knowledge needed to prepare drugs, that have a profile of activities specifically directed to the desired therapeutic effect. Numerous drugs in clinical use have one or more asymmetric centers. In many instances only racemic mixtures of the different stereoisomers are available. Since the intended effectiveness is often limited to one stereoisomer, the mixtures contain stereoisomers that hardly contribute to the effect. The presence of therapeutic ineffective stereoisomers can influence the profile of action as compared to that of the therapeutic effective isomer alone.

Let us take an example of a racemic drug mixture containing two enantiomers, one of which mediates the desired biological activity. Differences in bioavailability, because of differences in absorption and/or first pass effect, differences in the volume of distribution due to a more or less stereoselective transport across membranes and/or binding to proteins and variations in biological half-life because of differences in biotransformation, and/or excretion, may exist for the two enantiomers. Furthermore, besides the desired biological activity, which in our example is stereoselective, the drug will cause a variety of other effects (toxic or non-toxic), that may obey to different rules of stereoselectivity. The picture may become even more complicated, when taken into account the possibilities of biological active metabolites, that may be stereoselective due to the inherent structure of the parent molecule or to the creation of a new asymmetric center during metabolism. The ultimate result may be that the mixture of the two enantiomers has qualitatively and/or quantitatively different actions than the active constituant.

At best the mixture has half the potency of the active substance, without an increase in known, undesirable effects. But even in this case one has to consider the safe load of the patient and the environment since no chemical substance is inert. When the profile of action is changed, effects unrelated to the desired action are increased, toxicity is enhanced, etc., one should prefer the active stereoisomer over the racemic mixture. Only when the non-active enantiomer is more or less antagonizing the unwanted effect of the active enantiomer, the mixture may be favourable.

However, we have to consider such a mixture as a combination of two drugs, which may have consequences for registration of that mixture for therapeutic use. Although I am aware of the fact that this viewpoint is primarily based on theoretical considerations, it may have important practical consequences in the near future. Ideally, already in an early stage of drug development, research should be directed towards selection of the most active stereoisomer and this compound should be used for further manufacturing. In this respect it is worthwhile to mention that in the peptide field, where at present a variety of new drugs are being developed, not only for classical endocrine diseases, but also for different psychopathological and psychiatric disorders, stereoselectivity is a common phenomenon. In fact, the various enantiomers of peptide molecules frequently have different or even antagonistic biological effects and are considered as separate substances, When this principle is adapted for all biological active molecules, one may expect the development of new drugs for therapeutic use in the next decades, that are highly active, more selective and less toxic.

Origin and Basis of Stereoselectivity in Biology

A.G. Rauws

Abstract

Stereoselectivity is the extent to which an enzyme or other macromolecule or macromolecular structure (antibody or receptor) exhibits affinity towards one molecule of a pair of stereoisomers in comparison with and in contrast to the other isomer.

The necessity of asymmetrical building blocs for enzymes and other stereoselective macromolecules is argued. A review is given of theories and experiments on spontaneous production of excess enantiomer.

The importance of kinetics and thermodynamics of open systems in understanding origin and maintenance of molecular asymmetry is stressed.

The honour and pleasure of delivering a chapter in this book I owe, I guess, to the excellent memory of the editors, who remembered that almost twenty years ago I have written in the Dutch language a critical review of then current theories about prebiotic organic chemistry[1].

In this chapter I will try to give you an idea of how molecular asymmetry might have originated, basing myself on the evolutionary paradigm, which not only embraces biology, but also cosmology and geology. In this view biology is a later and local phase in cosmological evolution.

To my astonishment and embarrassment I have not been able to find a definition of stereoselectivity. So I made this one myself: Stereoselectivity is the extent to which an enzyme or other macromolecule or macromolecular structure (antibody or receptor) exhibits affinity towards one molecule of a pair of stereo-isomers in comparison with and in contrast to the other. Lehmann[2,3] has expressed this in a mathematical form: the ratio of activity or affinity of the better fitting enantiomer (eutomer) to that of the less fitting enantiomer (distomer) is defined 'eudismic ratio'. From this he derives the 'eudismic affinity quotient' as a quantitative measure of stereoselectivity (Fig. 1). In the following we will concern ourselves with optical isomers, although this definition is also applicable to other forms of stereo-isomery. However this difference exist: enantiomers have the same energy content - or should I say approximately the same? - whereas other pairs of geometrical isomers may exhibit appreciable differences in energy content. So the speculations in this lecture will be focused upon the origin of molecular assymmetry or enantioselectivity.

In our daily life we are accustomed to see and guess symmetry around us. The architecture of the buildings which dominate the scenery of our largely urban civilisation imprints a bias for symmetry, which modern art is scarcely able to neutralize. The strength of this bias is such that we tend to overlook the fact, that inside a building, behind the neatly symmetrical façade an array of rooms with different forms and different functions produces a quite unsymmetrical picture. In the same way our nearly symmetrical body hides the asymmetrical arrangement of our intestines (Figs. 2 and 3). Function is less symmetrical than façade, and increasing functional specialization entails increasing asymmetry.

Eutomer : enantiomer with higher affinity or activity

Distomer: enantiomer with lower affinity or activity

$$\frac{Aff_{Eu}}{Aff_{Dis}} = \text{eudismic ratio}$$

$$\text{Log } Aff_{Eu} - \text{Log } Aff_{Dis} = EI = \text{eudismic index}$$

In a series of agonists/antagonists:

$$EI = a + b \text{ Log } Aff_{Eu}$$

$$b = EAQ = \text{eudismic affinity quotient}$$
$$= \text{measure of stereoselectivity}$$

Figure 1. *Nomenclature and definitions in stereoselectivity*

Let us now turn to biochemistry and take an enzyme as an example. The intricate web of chemical reactions in living systems requires a highly envolved selectivity and even specificity of the tools involved: the enzymes. Here also increased functional selectivity asks for increasing asymmetry. Why? Most people here will accept this as plausible. It might however be illuminating to try and derive this thesis stepwise. The function of an enzyme is to lower the energy barrier opposing a certain reaction at a certain substrate. To achieve this the substrate molecule has for example to be bound to the enzyme in a conformation making it less stable, meanwhile coming into contact with regions of the enzyme molecule which will transiently take part in the ensuing reaction.

In the first place an enzyme exhibits reaction specificity, which means that, to be able to catalyze the reaction in question the spatial arrangement of the interacting groups has to be reproducible within rather narrow margins, allowing thereby for the conformational changes the protein molecule has to make to accomodate the substrate molecule at the active site (induced fit)[4]. The tool has to exhibit very little play, as a mechanic would say. As an example one might mention the interaction of lysozyme and an hexasaccharide substrate analogue[5].

In the second place the higher the selectivity towards the substrate, the more other regions of the polypeptide chain will have to conform to requirements of conformational reproducibility.

In a review on the origin of optical activity George Wald made the following thoughtful observation[6]: 'Only the fact that chemistry is learned from the plane surfaces of paper and blackboard makes such selectivity seem strange. We tend to think of optical isomers as very much alike, but in fact they represent profound differences in *shape*; and, in the types of reactions upon which life depends, involving the ceaseless, intimate fitting together of molecular surfaces, shape is all-important.'. Indeed in biochemistry L-alanine should be considered as quite an other compound than D-alanine, and the replacement of L-alanine by D-alanine in a protein molecule is a much more drastic procedure in three dimensions than it looks like in two dimensions. The right handed α-helix regions along the poly-

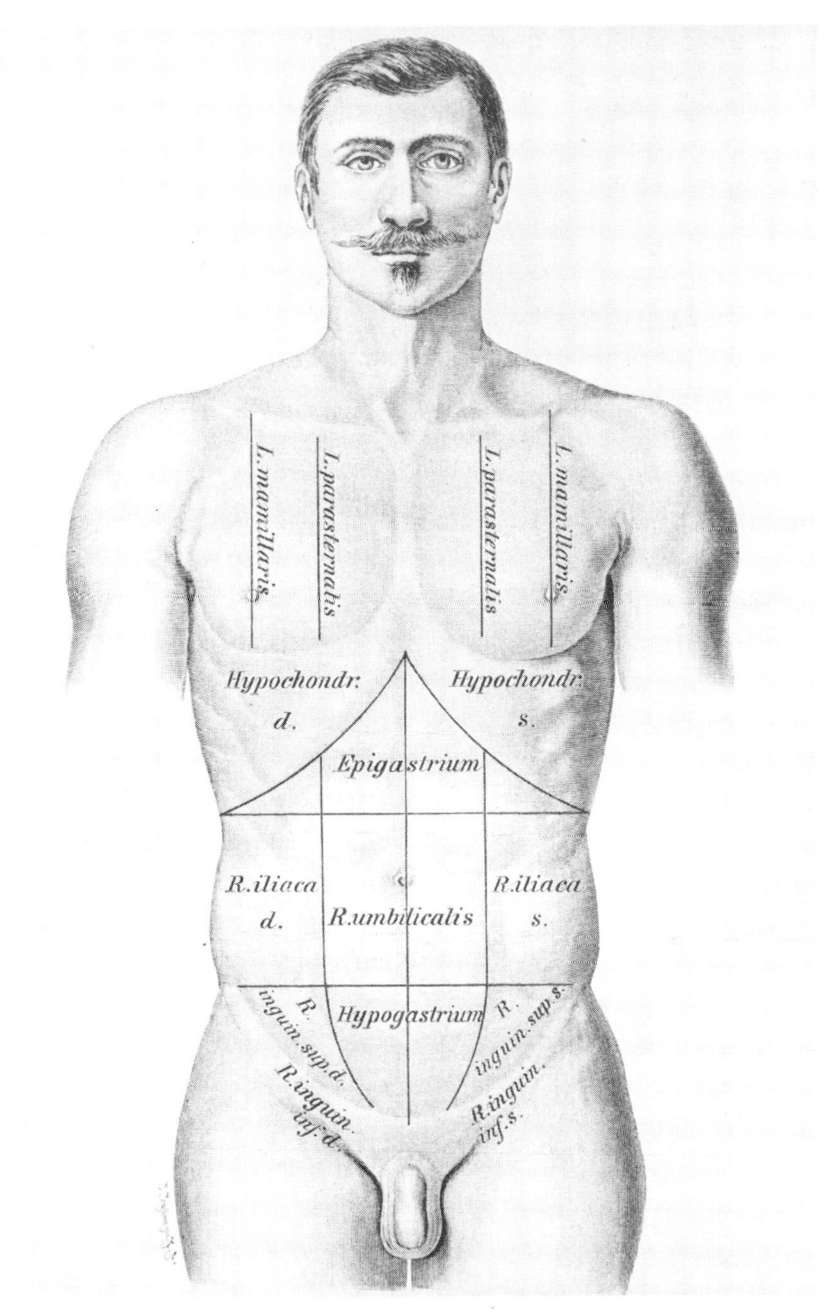

Figure 2. *External symmetry of the human body*

peptide chain form the stakes and the disulfide bridges or other intra-chain interactions form the connections between the stakes which together make the protein to a structure with a reproducible conformation (or a limited set of conformations with a view to induced fit). If we grow a mixed chain of L- and D-amino acids on a right handed α-helical primer it grows further as such, but its growth rate is less and it is definitely less stable than the same α-helix exclusively consisting of L-amino acids (Fig. 4)[7,8]. Moreover we would obtain a population of structurally different poly-

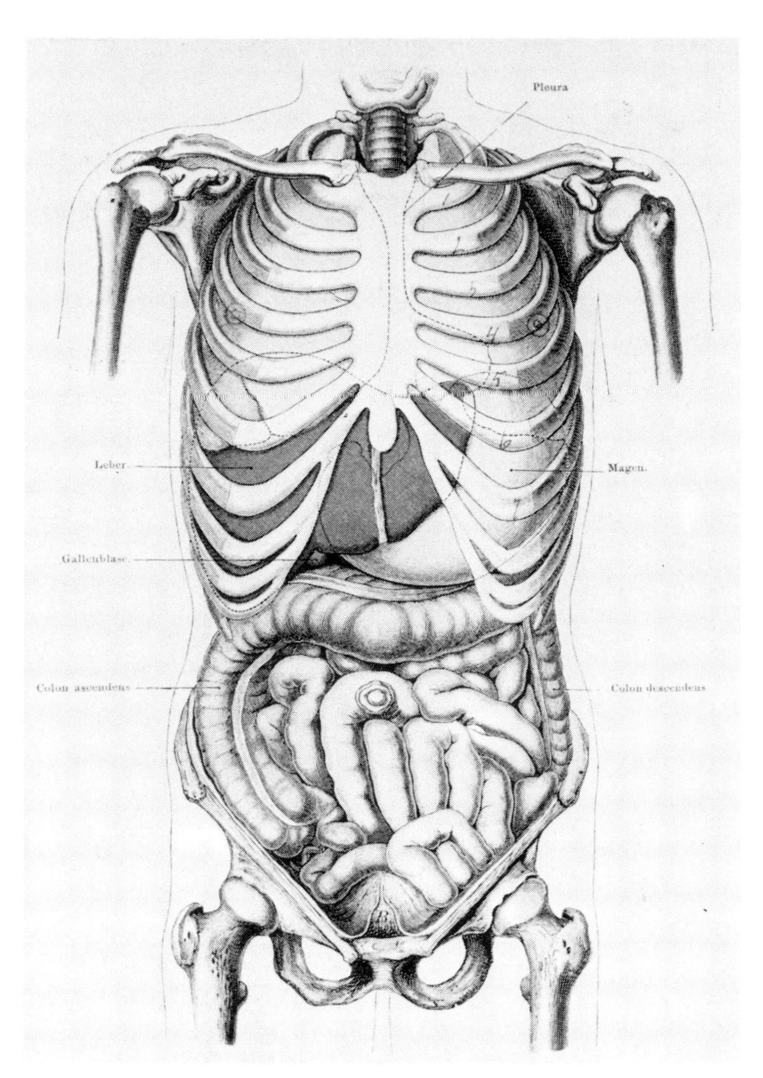

Figure 3. *Internal assymmetry of the human body*

peptides, even if it were possible to maintain the same amino acid sequence in each chain. Along with every active enzyme molecule a large population of inactive stereo-isomers would be synthetized, a rather disturbing and energetically disadvantageous form of 'chemical noise', which, as we know, does not exist.

The elucidation of the catalytic mechanisms of enzymes like lysozyme, carboxypeptidase A, papain and others, with the assignment of roles in the catalytic process to side chains of individual amino acids clearly shows to us the all-important point of conformational reproducibility. Both enzyme and substrate have to satisfy structural and conformational requirements (Fig. 5)[9]. To achieve this reproducibility an intramolecular stereoselectivity is required. This internal stereoselectivity can only be obtained with asymmetric building blocs. With polyglycine we cannot build the α-helix, with hypothetical symmetric amino acids, carrying two identical side chains at the α-C atom we cannot build the α-helix either: there is no place for additonal groups.

The preceding line of thought brings me to the conclusion, that the preponder-

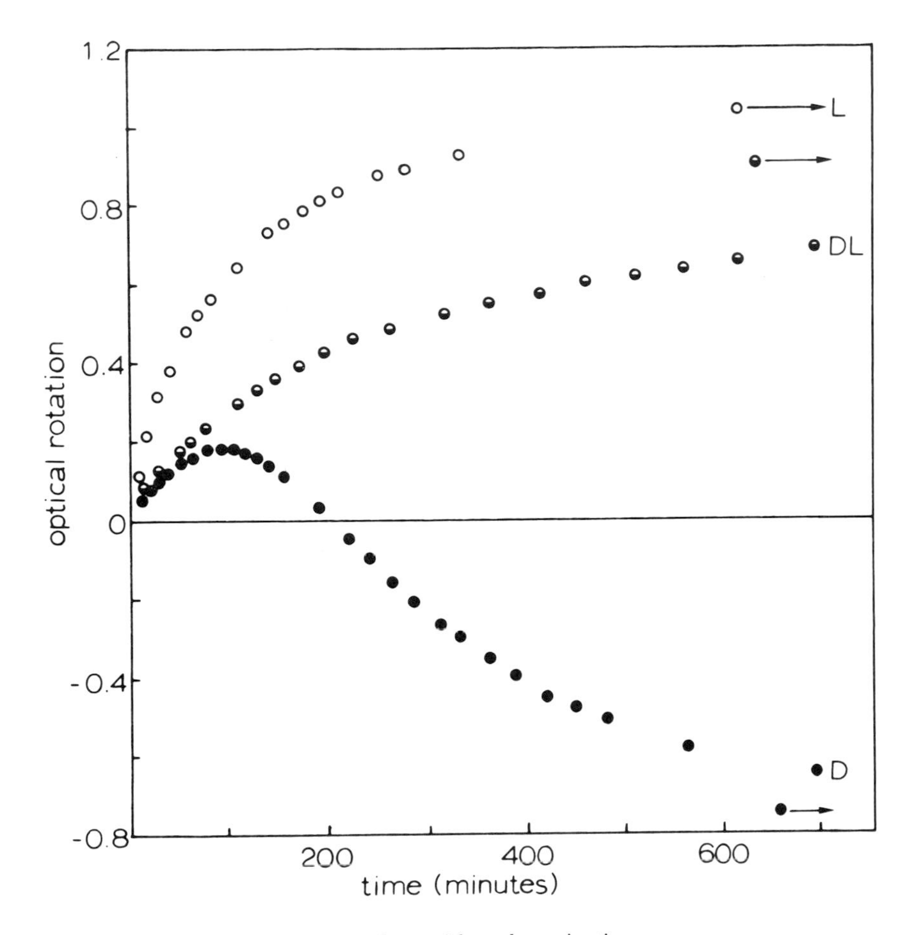

Figure 4. *Helix formation in amino acids polymerization*
L : Complete righthanded α-helical conformation with L-amino acids
DL: Mixed chain D- and L-amino acids with less stable and incomplete helical conformation
D : Growth of D-amino acid chain on L-primer. Reversal of helical handedness to lefthanded α-helix
Reproduced with permission from ref. 7.

ance of one enantiomer over the other in amino acid, nucleic acid or carbohydrate metabolism should have been established before the origin of life as we know it now. In which phase would this selection then have taken place? Let us review the current theories about the origin of molecular asymmetry.

An essential requirement for any proposed mechanism for the production of excess enantiomer will be, that its rate is not only larger than the racemization rate in the same environment, but that it does not decline when the concentration of the other – 'underdog' – enantiomer decreases towards zero.

Much experimental work has been done in the past on the subject of 'spontaneous' asymmetric synthesis. In many cases the investigators had unconsciously built in their own little Maxwell demon into their experimental design. Catalysis at surfaces of optically active crystals, circularly polarized radiation and many other kinds of asymmetric physical influences have been summoned to tip the balance of racemization out of equilibrium. Bonner, working himself in the field, has written

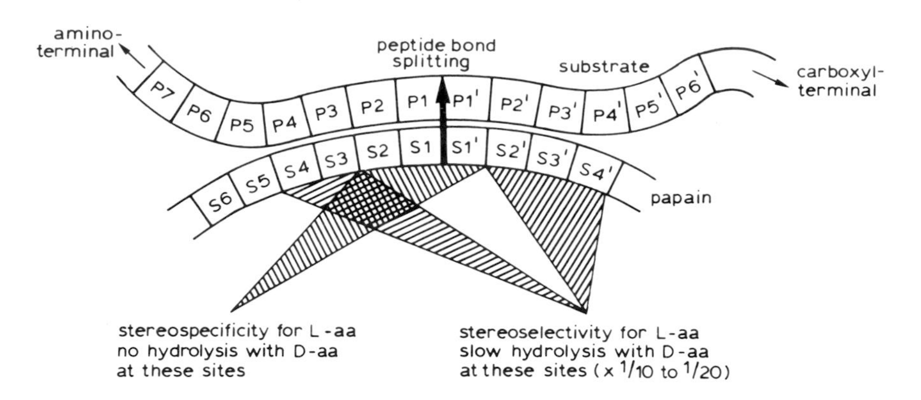

Figure 5. *Active site of papain and its subsites*

some interesting reviews on the subject[10,11].

In 1956 the theoretical physicists Lee and Yang postulated that parity is not conserved in β-decay, and Wu demonstrated this in an elegant experiment. Since then the notion of an inherent asymmetry in our Universe has been disseminated also in the biological sciences. Ulbricht and Vester[12] have shown the theoretical possibility of coupling between asymmetry at the level of elementary physical particles and at the level of molecules. They stressed, that this did not necessarily mean that a causal connection did exist in nature. Garay, Bonner and others have carried out experiments with β-irradiation and 'Bremsstrahlung'[13-16]. The effect - production of escess enantiomer - was however minimal and only extremely careful design and execution of the experiments enabled the investigator to exclude chance production of excess enantiomer[13].

Whatever may be the ultimate cause of the production of excess enantiomer: the infinitesimal excess produced has to start some amplification mechanism which leads to the production in the end of optically almost pure enantiomer. As early as 1953 Frank designed a system in which the enantiomers autocatalytically increased their numbers, meanwhile inhibiting the production of the other enantiomer (Fig. 6)[17]. Moreover he exactly predicted the properties of this system in a larger - primordial - environment. His work remained almost unnoticed. The design of Frank gives an impression to be rather artificial. However already during the early forties Havinga, in the wartime Netherlands, carried out experiments producing spontaneously optically active substances. The results were published after the war in a more accessible form[18].

Havinga used a system closely analogous to that of Frank. He needed a substance not crystallizing as a racemate, but only in the D- or the L-form, exhibiting rapid racemization and rapid crystal growth but with a low rate of nucleation. A substance satisfying these requirements was methyl-ethyl-allyl-aniliniumiodide in chloroform solution (Fig. 7). Supersaturated solutions of this substance, prepared under - then extreme - precautions to exclude crystallization nuclei often did not crystallize in a year. Other showed formation of almost pure L- or D-crystals. In some cases a conglomerate of L- and D-forms was found.

The system described above, although successful, cannot be a model for what once will have happened in a later phase of prebiotic evolution, because it is a closed one. We have to look for open systems, where an interplay of catalyzing and inhibit-

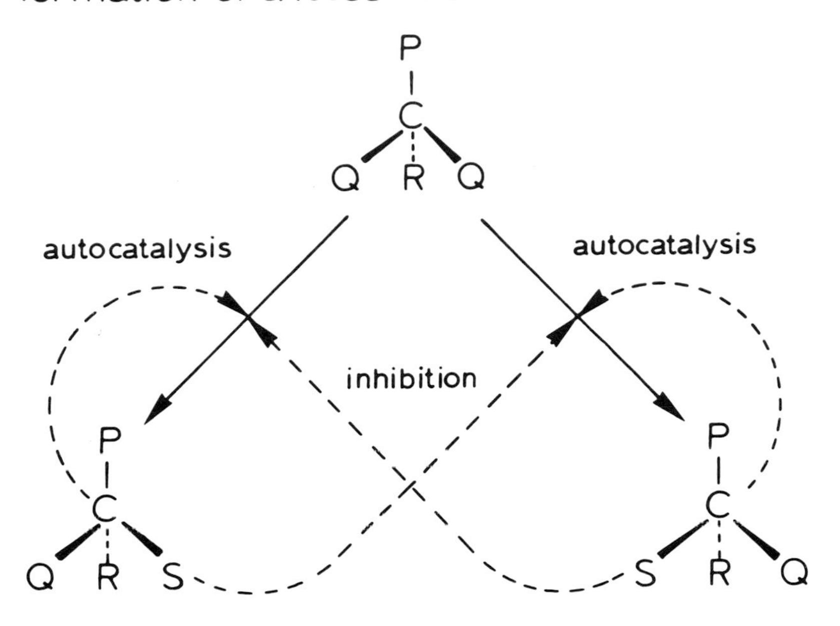

Figure 6. *Labile equilibrium system for spontaneous formation of excess enantiomer*

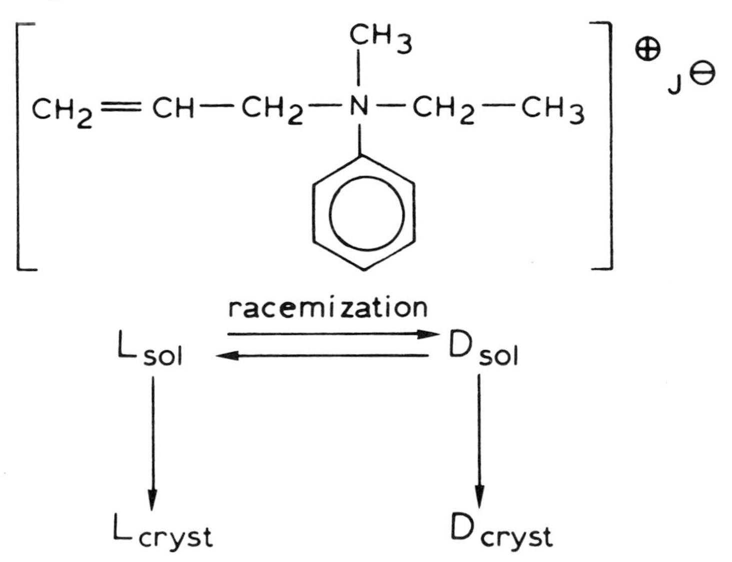

Figure 7. *Labile equilibrium system for spontaneous formation of excess enantiomer through crystallization*

ing actions may displace the steady-state from the racemic level to one or the other of the enantiomeric levels without violating the second law of thermodynamics. These systems have to obey the laws of non-equilibrium thermodynamics to prevent them falling back to equilibrium. Decker, building on the foundations laid by Frank, has shown us some of the possible kinetic constellations which may generate and amplify excess enantiomer in open systems. In a series of publications he has expounded the basic idea to considerable detail[19,20,21].

There are two different ways of producing excess enantiomer: selective production from a non-asymmetric precursor and selective degradation in a racemic mixture (Figs. 8 and 9). In both types the competitive efficiency may be increased by increasing the molarity of the autocatalytic reaction: hypercompetitive system, or by an additional mutually inhibiting mechanism: aggressive system.

Productive systems

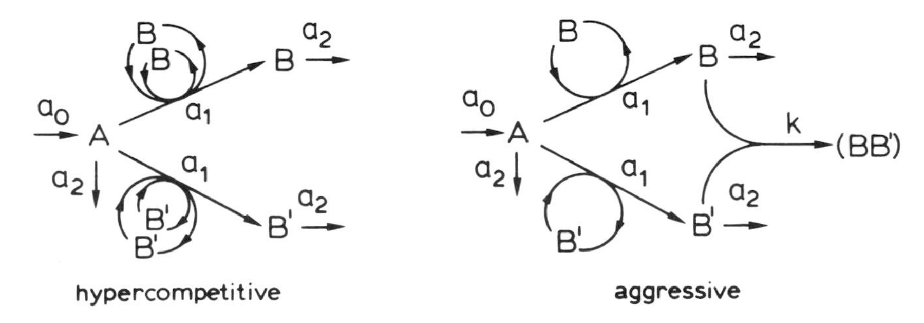

Figure 8. *Reaction schemes of hypothetical enantiomer producing systems From: P. Decker*[21]

Racemate resolving systems

hypercompetitive

aggressive

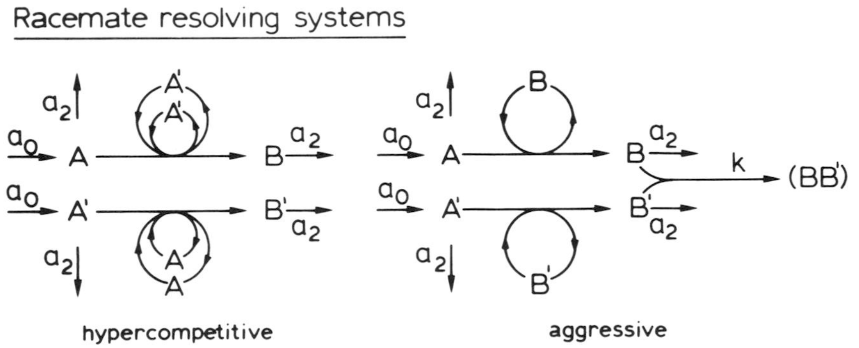

Figure 9. *Reaction schemes of hypothetical racemate resolving systems From: P. Decker*[21]

If then in a homogenous racemic system a domain arises with excess enantiomer, such a domain will grow, until it meets a domain with excess of the opposite enantiomer. An unstable frontier will then originate which will shift depending on the spatial orientation of the domains (Fig. 10). At a curved frontier the domain at the convex side will, for kinetic reasons, grow at the expense of the one at the concave side. Only if two domains are connected by a narrow channel a stable frontier will result.

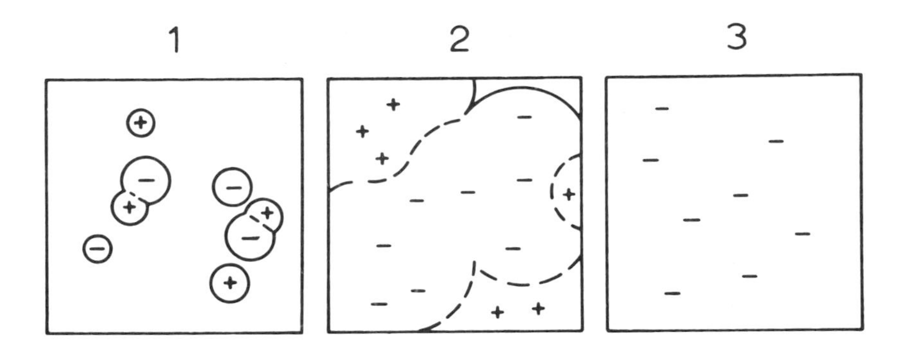

Territorial evolution of incompatible systems

Figure 10. *Territorial evolution of incompatible systems*

Unfortunately until now these and other systems[22,23] arc largely hypothetic. It would be an important development if it were possible to build a reactor carrying out a realistic prebiotic asymmetric synthesis, using an inflow of a probable abiogenic organic compound. This is the kind of system Decker calls a bioid. If the antipode selection by polymerization[24] could be incorporated in a steady-state system, this would be a large advance in the right direction.

In a broader context Manfred Eigen has combined thermodynamics of irreversible systems, information theory and biochemistry to a theory of self-organization and evolution on the molecular level[25]. It is impossible to elaborate his theories within the frame of this lecture, but I mention his important work in this field, because the origin of enantioselectivity is just one process in a large group of prebiotic evolutionary processes, which will have occurred simultaneously in a distant past. In a more general monography Eigen has expounded his views for the persevering layman[26].

I have tried in this introduction to give a brief review of origin and basis of stereoselectivity, especially enantioselectivity. In 1962 I wrote that, after much work on abiogenic synthetic pathways, progress in prebiotic chemistry had to come from thermodynamics and kinetics of steadystate processes[1]. Revisiting after many years the scene of prebiotic theory and speculation I was pleasantly surprised to see, how important the thermodynamics of the steady-state, or of irreversible processes have become in this context. But at the same time I feel somewhat frustrated that the subject has escaped the grasp of the common biochemist and has become the hunting-ground of the theoretical chemists[25,27]. Well, I wish them good hunting and look forward to the trophies they will bring back, showing us a deeper perspective on the origin and nature of life.

References

1. A.G. Rauws, Chem. Weekbl. *64*, 361 (1964).
2. P.A. Lehmann, J.F. Rodrigues de Miranda and E.J. Ariëns, Progr. Drug Res. *20,* 101 (1976).
3. P.A. Lehmann, Trends Pharmacol. Sci *3,* 103 (1982)
4. D.E. Koshland, Sci. Amer. *229* (4), 52, (1973).
5. M.O. Dayhoff and R.V. Eck, Atlas of Protein Sequence and Structure. National Biomedical Research Foundation, Silverspring Md., 1973.
6. G. Wald, Ann. N.Y. Acad. Sci. *69,* 352 (1957).
7. P. Doty and R.D. Lundberg, J. Am. Chem. Soc. *78,* 4810 (1956).

8. E.R. Blout, P. Doty and J.T. Yang, J. Am. Chem. Soc. *79,* 749 (1957).
9. J. Drenth, J.N. Jansonius, R. Koekoek and B.G. Wolthers, Adv. Prot. Chem. *25,* 79 (1971).
10. W.A. Bonner, In 'Exobiology' (C. Ponnamperuma, ed.), North-Holland, Amsterdam, 1972.
11. W.A. Bonner, In: 'Origins of Optical Activity in Nature (D.C. Walker, ed.) Elsevier, Amsterdam, 1979.
12. T.L.V. Ulbricht and F. Vester, Tetrahedron *18,* 629 (1962).
13. W.A. Bonner, M.A. van Dort, M.R. Yearian, Nature *258,* 419 (1975).
14. A.S. Garay, Nature *219,* 338 (1968).
15. T.L. Kovacs and A.S. Garay, Nature *254,* 538 (1975).
16. A.S. Garay, L. Kesthelyi, I. Demeter and P. Hrasko, Nature *250,* 332 (1974).
17. F.C. Frank, Biochim. Biophys. Acta *11,* 459 (1953).
18. E. Havinga, Biochim. Biophys. Acta *13,* 171 (1954).
19. P. Decker, Nature New Biol. *241,* 72 (1973).
20. P. Decker, Orig. Life *6,* 211 (1975).
21. P. Decker, In: Origins of Optical Activity in Nature (D.C. Walker, ed.), Elsevier, Amsterdam, 1979.
22. K. Wagener, J. Mol. Evol. *4,* 77 (1974).
23. F.F. Seelig and O.E. Rössler, Z. Naturforsch. *27b,* 1441 (1972).
24. W. Thiemann and W. Darge, Orig. Life *5,* 263 (1974).
25. M. Eigen, Naturwissenschaften *58,* 465 (1971).
26. M. Eigen and R. Winkler, Das Spiel. Naturgesetze steuern den Zufall. Piper, München, 1975.
27. I. Prigogine, G. Nicolis and A. Babloyantz, Physics Today *25,* 38 (1972).

Stereoselectivity of Bioactive Agents: General Aspects

E.J. Ariëns

Abstract

The article has as a main goal the exemplification of various fundamental aspects of sterical structure in relation to biological action.

Implications of stereoselectivity of bioactive agents are discussed on basis of the concept of a three-point interaction between the bioactive agent and its target molecule, receptor or enzyme. Attention is particularly focussed on enantiomers characterized by their mirror image relationship.

The significance of stereoisomerism in daily practice, particularly as far as the use of mixtures of isomers as therapeutics, pesticides, etc. is concerned, is emphasized. Stereoisomers often greatly differ in bioactivity. Thus application of an active agent, the active isomer - the eutomer -, goes often and sometimes unknowingly, hand in hand with the application of an equal or even larger dose of an inactive agent, the inactive isomer - the distomer -. The latter does not contribute to the effect aimed at, but undoubtedly brings about a rejectable toxicological risk, to be avoided as far as possible.

Intriguing aspects of stereoselectivity in bioactivity are brought to attention. Examples are: the behaviour of one isomer as an agonist and the other as its competitive antagonist and the switch in sterical configuration in the conversion by chemical manipulation of the α-adrenergic norepinephrine to its α-adrenergic blocking aralkyl derivative. In relation therewith the separation of the 'combined' α-adrenergic blocking and β-adrenergic blocking action of 'the drug' labetalol in two of its isomers, the SR- and the RR-isomer, is noticed.

Stereoselectivity as a tool in the analysis of mechanisms of action, such as the role of various moieties in the molecule in its interaction with the receptor site or the active site on an enzyme, is illustrated. On this basis the concept of accessory binding areas on the receptor molecule involved in the action of various types of competitive antagonists, of particular importance for the understanding of structure action relationship, is elucidated.

Various aspects of the relationship between the activity ratio of stereoisomers - the eudismic ratio - and the activity of the more active isomer, the eutomer, are clarified.

The relation between sterical structure and selectivity in action is exemplified on basis of a number of diphenhydramine derivatives. Finally the significance of 'sterical structure' of position isomers and of chemical homologues for bioactivity is discussed.

Introduction

An intriguing aspect of biologically active agents, such as neurotransmitters, hormones, allosteric modulators, drugs, etc. – messenger molecules – is their discriminatory capacity with regard to the molecular sites of action, the specific receptors. No doubt, this capacity is based on a chemical complementarity between the messenger molecule or drug and the specific receptor site, the site of binding on the receptor molecule. In this complementarity, besides the physicochemical characteristics of the groups (ligands), which participate in the interaction, also their spatial arrangement, their sterical configuration, is essentially involved. The stereospecificity in the interaction can be counted for on basis of as few as three ligands, binding groups, in the molecule[1,2]. This even holds true for the high selectivity in the interaction between antibody and antigen. The three hypervariable regions in the light chains of the antibody are the determinants here[3]. The contributions to the binding energy obtained from the interaction of pairs of corresponding groups in the bioactive agent and receptor site (Table 1)[4] show that participation of a triade of pairs can generate binding energies in the order of 12-17 $kcalM^{-1}$ corresponding to binding constants in the order of 10^{-9} to $10^{-12}M^2$.

Table 1. *Chemical bonds in drug-receptor interaction*

Type of bond	Binding energy (kJ/mol)	Action radius
Hydrophobic	–4.0	small
Ionic	–2.0	large
Dipole-Dipole	–(4-30)	small
Hydrogen bridge	–(4-30)	small
Charge transfer	–(4-30)	small
Covalent	–(160-450)	

Action radius: 0.2-0.4 nm (small), 0.5-1.0 nm (large) after Eberlein[4].

Inasmuch as the carbon atom is the basis for biochemistry, stereochemistry is inherent to biological processes. The four valencies of carbon imply that, if to one carbon four different groups are attached, the spatial arrangement results in a center of asymmetry and occurrence of two stereoisomers, enantiomers characterized by their mirror image relationship. Although exceptionally, such a stereoisomerism is also possible on basis of nitrogen and phosphor atoms. In the simplest case, in solution, the isomers only differ in the spatial arrangement of the four different groups in the isomers. The corresponding groups in the mirror image do not differ in their chemical environment. Thus in solution such isomers will not differ in their NMR-spectra. Once interaction with a specific, that is 'stereoselective', receptor site takes place, differences between the stereoisomers will show up. They then also differ in their NMR-spectra from which information on the participation of particular groups in the interaction between the bioactive agent and its receptor site can be obtained.

If more than one center of asymmetry is located in one bioactive molecule (the case of diastereomers) and in other cases of geometric isomers such as: cis-trans-isomers, chair and boat configurations, epimers, isomerism on basis of intramolecular sterical hinderance, etc. as a result of differences in the intramolecular relationship between the various groups in the isomers, these will differ physico-chemically. Therefore, as a rule such isomers are more easily separated than the enantiomers of compounds with a single center of asymmetry. Separation of the latter is still a laborious task. This makes understandable that in practice, drugs, insecticides, weed-killers, and in general industrial products with bioactivity, often are marketed as racemic mixtures. Products from biological origin, such as particular hormones, antibiotics, etc. usually, by nature, are obtained in stereospecific form.

Practical implications of stereoisomerism

In daily practice, often unknowingly or unaware, in the supposition that just 'one' compound is applied, mixtures – 50:50 or even 25:25:25:25 – of compounds (stereoisomers) are applied. Since in biology stereoselectivity is more rule than exception, often only one of the components in the mixture appears to be truly active. Strictly taken the inactive isomers in the mixture should be regarded then as impurities, not contributing to the effect aimed at. They potentially contribute, however, to the unwanted effects, side effects, and to chemical pollution – be it of the milieu interne of man and animals – or of the environment in general. Taken into account the growing apprehension on chemical pollution, one has to be well aware of this situation. In medicine, there apparently is not too much concern about applying along with e.g. 50 mg of the compound with the desired action, 50 mg of a second compound with no contribution to the desired effect although, potentially contributing to the side effects. For certain types of therapeutics, such as β-adrenergic agents, β-adrenergic blockers, anti-epileptics and oral anticoagulants, up to 90% of the 'compounds' are in fact racemic mixtures. For antihistamines and local anaesthetics this holds true for about 50% while on the whole it concerns 10 to 15% of the drugs. In some cases the differences in activity of the enantiomers are well established, in many cases no information is available.

In the study of bioactive agents, it is preferable if not a prerequisite, to use pure agents, that is compounds with as little as possible impurities. Mixtures of compounds may be an object of study, particularly if one wants to gain information on the interaction between the compounds in the mixture. The use of mixtures of compounds, without being aware of it, is not acceptable. Too often, even without noticing, in the literature data on mixtures of stereoisomers are presented as if they concerned one compound. In those cases that such mixtures are used, because of the non-availability of the stereoisomers, one at least should mention explicitely that the data presented concern a mixture of compounds which possibly differ in their bioactivity. The implications of the use of mixtures of active and inactive isomers become more clear if one considers the application of, for instance, pesticides. What about the presumably economically and environmentally acceptable spraying of e.g. 500 kg of an active agent, the active isomer, automatically going hand in hand with spraying of 500 kg of a second chemical, the inactive isomer, not contributing to the effect but undoubtedly to be regarded as a risk?

A number of the pesticidal organophosphates have a center of asymmetry. This may be located on a carbon atom like in dialiphos and malathion or on the phosphor atom like in methamidophos and trichloronate. The compound soman has two chiral centers, one on a carbon and one on a phosphor atom, which implies four stereoisomers. Differences in the capacity to inhibit acetylcholine-esterase and therewith differences in the toxicity of stereoisomers of such organophosphates are well established, particularly for the agents with the phosphor atom as a chiral center[5-7] (Fig. 1).

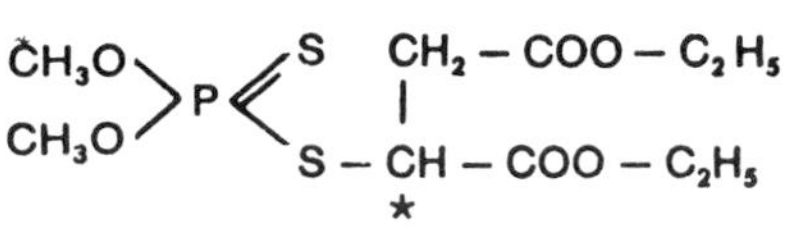

Malathion

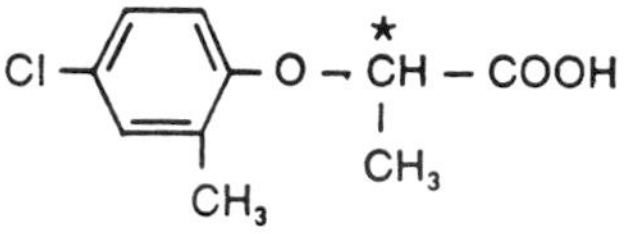

Mecoprop

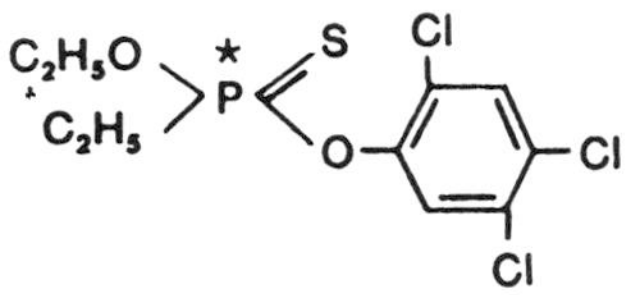

Trichloronate

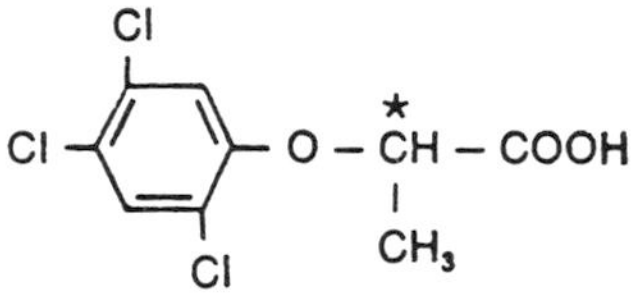

Phenoprop

Figure 1. *Insecticides and herbicides applied as mixtures of stereoisomers. Note: both the insecticide trichloronate and the herbicide phenoprop are derivatives of trichlorophenol, known to be a source for tetrachlorodibenzo-p-dioxine (TCDD), an extremely toxic agent (Seveso). This risk is inherent to both the active and the inactive isomers of the pesticides concerned.*

For herbicidal agents, e.g. defoliants, too, large differences in the activity of antipodes are observed. This is the case, for instance, for the ring-substituted α-phenoxy-propionic acids such as mecoprop (CMPP) and phenoprop (2,4,5-TP) of which the (+)-form is the active isomer[8] (fig. 1). Without calling the situation prohibitive, and requiring the use of only the active isomers, the problem should at least be recognized. Avoidance of unnecessary application of chemicals, be it as drugs, pesticides or whatever should count heavily both in the development of new agents and in the efforts to optimize safety and therewith acceptability of chemicals in daily use. Chemistry, undoubtedly, has essentially contributed to health and the standard of life in many parts of the world. The risks, however, have been underestimated, such that rather 'wild' chemicals were accepted. Time is ripe for true 'domestication' of chemistry; that is adaptation of chemistry to safe intimate association with and advantage of men, subjected to his control.

Stereoselectivity in bioactivity

If in a set of two stereoisomers, there is a difference in activity, the more active isomer may be termed the eutomer and the less active isomer the distomer, this regardless of their absolute configuration but only with regard to a particular biological action. The degree of stereospecificity, that is the ratio of the activities (affinities, potencies, etc.), may be termed the eudismic ratio[9,10,26].

Bioactive agents often show more than one type of action. The question then arises whether the various components in the action are based on one and the same mechanism or whether different mechanisms are involved. In case stereoselectivity – the ratio for the potencies of the isomers, the eusdismic ratio – differs for the different components in the action, a difference in the mechanism of action, a difference in pharmacodynamics, may be the cause. However, also differences related to stereospecificity in pharmacokinetics such as in metabolic conversion, transport or storage may be involved. Studies on isolated tissues and binding studies on tissue preparations give more direct information in this respect. Table 2[11] gives an example of the differentiation between components in the action of bioactive agents on basis of a study of the stereoisomers. The β-adrenergic blocking action is clearly stereoselective which is not the case for the local anaesthetic action. This indicates difference in mechanism of action involved.

Table 2. *The ratios for the activities of l- and d-isomers (l/d) for β-adrenergic blocking agents with regard to different components in their action*

	Ratio for β-adrenergic blocking activity	Ratio for local-anaesthetic activity
DCI	40	1
Pronetalol	100	1
Propranolol	100	1
Aptin	>100	1

The β-adrenergic blocking activity is clearly dependent on the sterical configuration which is not the case for the local-anaesthetic activity. This indicates that different molecular mechanisms are involved[11].

In a study of structure-action relationship in a series of norepinephrine derivatives with gradually enlarged substituents on the amino group it was observed that with the increase in the substituent, particularly if aralkyl groups are introduced, the β-adrenergic action of norepinephrine is maintained, it even increases. The α-adrenergic action, however, is lost and in fact converted to an α-adrenergic blocking action[12,27].

Comparison of the optical isomers of the derivatives, particularly of the aralkyl substituted derivatives, shows that the β-adrenergic action is present in predominantly the isomer with the natural configuration (that of 1-norepinephrine), which is in fact as expected[13]. The α-adrenergic blocking action, however, turns out to be present predominantly in the other isomer. This implies a switch from the configuration of 1-norepinephrine essential for the α-adrenergic action to that of d-norepinephrine for the α-adrenergic blocking action (Table 3)[11,12,29]. So the 1-norepinephrine derivative is the eutomer for the β-adrenergic action and the d-norepinephrine derivative the eutomer for the α-adrenergic blocking action. These are both components in the action of the racemic mixture. This emphasizes the relativity of the indication eutomer and distomer.

Of paticular interest in this relation are the isopropyl derivatives of 1- and d-norepinephrine, the racemic mixture is known as isoprenaline. As shown in Table 3, here the isomer with the configuration of 1-norepinephrine behaves as an α-adrenergic while the other isomer behaves as an α-adrenergic blocking agent. The racemic mixture behaves as a partial agonist (Fig. 2)[11,12]. This manifests how one can be misled by relying on study of racemic mixtures.

The peculiar situation of two antipodes behaving as competitive antagonists is not unique. Various examples are found in the field of herbicides, namely among the

Table 3. *The relationship between sterical structure and activity ratios – eudismic ratios – for the optical isomers of N-alkyl and N-aralkyl substituted norepinephrine derivatives for their α-adrenergic action (vas deferens of the rat) and their β-adrenergic action (atrium guinea pig). The l-isomers have a configuration identical to that of l-norepinephrine, the d-isomers identical to that of d-norepineprhine.*

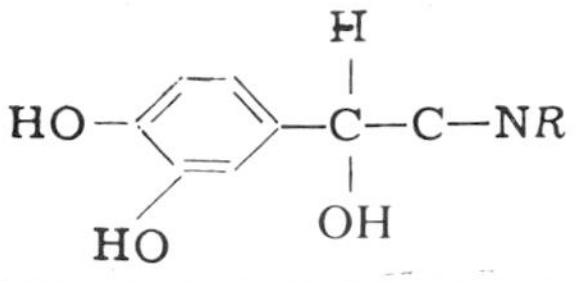 HO–⟨ring⟩(–HO)–C(H)(OH)–C–NR	α-adrenergic affinity-ratio l/d stimulant	blocker	β-adrenergic affinity-ratio l/d stimulant
−R = −H	4		20
−C	8		50
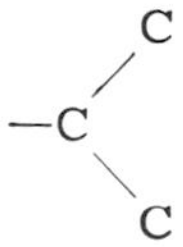 −C(C)(C)	$l_{mim.}$	$d_{lyt.}$	> 500
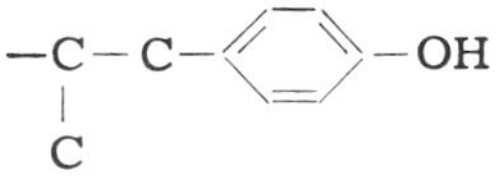 −C(C)−C−⟨ring⟩−OH		1/10	>1000

Note: with an increase in the size of the substituents introduced in norepinephrine a conversion of the α-adrenergic (α-mimetic) action into an α-adrenergic blocking (α-lytic) action takes place. This goes hand in hand with an inversion in the eudismic ratio (see also Fig. 2). The β-adrenergic action is maintained and even increases for the derivatives with larger substituents[12,29].

ring-substituted phenoxy-propionic-acids and α-2-naphthyloxy-propionic-acids[9].

In case of a large eudismic ratio like, for instance, for acetyl-β-methylcholine a factor 200, the racemic mixture has an apparent affinity about half that of the eutomer. A comparison of these isomers with acetylcholine itself learns that the apparent affinity of acetylcholine is practically equal to that of the active isomer of acetyl-β-methylcholine[14,15]. This indicates that in the distomer the β-methyl group causes sterical hindrance on the receptor site, while in the eutomer the β-methyl group does not disturb the complementarity to the receptor site.

For norepinephrine and epinephrine the apparent affinities of the distomers are practically equal to that of dopamine and epinine, respectively, the analogues lacking the alcoholic OH-group. The activity ratios are close to one. This indicates that in the eutomers this OH-group essentially contributes to the receptor binding, while in the distomers it is in a position away from the receptor[11–13].

As shown in Table 3, introduction of aralkyl substituents on the amino function of norepinephrine results in a switch from α-adrenergic to α-adrenergic blocking action. It is known that manipulation of the catechol moiety, particularly substitution on the ring combined with elimination of the catechol configuration results in a switch from β-adrenergic to β-adrenergic blocking action[11,16]. Combination of these two substitutions results in products with a combined α-adrenergic blocking and β-adrenergic blocking action. Labetalol is an example. It is introduced into the clinic as 'one drug' with both α-adrenergic blocking and β-adrenergic blocking capacity.

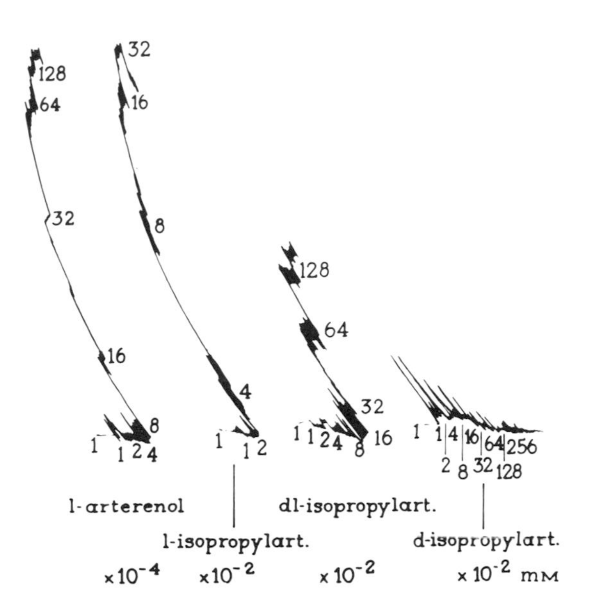

Figure 2a. *Cumulative log dose response curves for the optical isomers and the racemic mixture of isopropylnorepinephrine tested on the vas deferens of the rat. Note: the l-isomer acts as an α-adrenergic agonist, the d-isomer is apparently inactive but in fact acts as a competitive antagonist, an α-adrenergic blocking agent (Fig. 2b). As a consequence the racemic mixture, dl-isopropylnorepinephrine, behaves as a partial agonist.*[12,29]

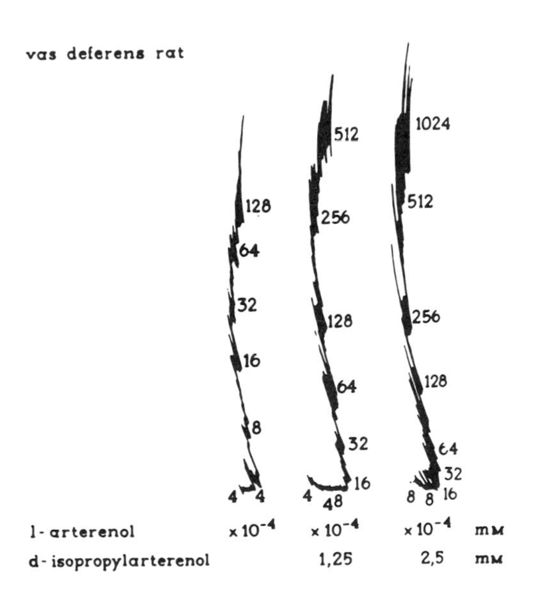

Figure 2b. *Cumulative log dose response curves for the α-adrenergic agonist l-norepinephrine in the presence of constant concentrations of d-isopropylnorepinephrine tested on the vas deferens of the rat.*
Note: d-isopropylnorepinephrine, inactive as such (Fig. 2a) behaves as a competitive antagonist, as an α-adrenergic blocking agent, with regard to the α-adrenergic action of l-norepinephrine[12,29].

One has to be well aware, however, of the fact that, taken into account the two centers of asymmetry in the compound, labetalol is composed of four stereoisomers (Fig. 3). Brittain et al.[17] studied the pharmacological properties of these isomers with the remarkable results shown in Table 4. It turns out that the eutomer for the α-adrenergic blocking activity has the SR-, that for the β-adrenergic blocking activity the RR-configuration. The SS-isomer and the RS-isomer are only poorly active. The β-adrenergic blocking activity requires for the alcoholic OH-group on the side chain a configuration identical to that of l-epinephrine[13]. Like in the example presented in table 3, here too there is a separation of the, this time α-adrenergic blocking and β-adrenergic blocking action. The α-adrenergic blocking action is located in an isomer with the configuration of d-epinephrine. So labetalol is definitely not 'one drug' with a combined α- and β-adrenergic blocking action.

RR-stereoisomer (AH 19501)

SS-stereoisomer (AH 19502)

SR-stereoisomer (AH 19504)

RS-stereoisomer (AH 19503)

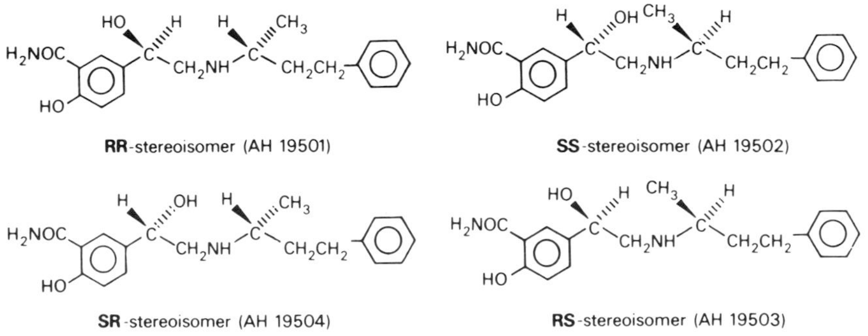

the individual stereoisomers of labetalol.

Brittain e.a. (17)

Figure 3.

Table 4. *The relationship between sterical structure and biological activity of labetalol, a drug with a combined β-adrenergic blocking and α-adrenergic blocking action.*
The α- and β-adrenergic blocking potencies of the stereoisomers are expressed as pA_2.

	Rabbit aortic strip/ (α_1-adrenoceptors)		Guinea-pig left atrium (β_1-adrenoceptors)		Guinea-pig tracheal strip (β_2-adrenoceptors)	
	n	pA_2	n	pA_2	n	pA_2
RR-isomer	3	5.87	6	8.26	5	8.52
SS-isomer	6	5.98	4	6.43	4	<6.0
RS-isomer	3	5.5	3	6.97	4	6.33
SR-isomer	8	7.18	4	6.37	4	<6.0

pA_2 calculated as described by Arunlakshana & Schild. R.T. Brittain, G.M. Drew & G.P. Levy[17].
The α-adrenoreceptor blocking potency rests predominantly in the SR-isomer, the β-adrenoreceptor blocking potency in the RR-isomer. The SS- and RS-isomers, 50% of the compound labetalol, can be regarded as practically inactive.

In this particular case, the presence of two, as far as information goes, practically inactive compounds, making up for 50% of the drug, does not seem to be of particular medical significance. In other cases, however, like for instance for the β-adrenergic blocking agent practalol, the therapeutic action is located in one isomer while the very serious side effects, 'the practalol syndrome', ascribed to the acetylanilide group in the molecule, may be assumed to be present in both isomers.

Another interesting aspect of the four stereoisomers in labetalol is that like for the example given in Table 2, the 'non-specific', negative dromotropic effect of the various isomers is not stereoselective (Table 5)[17].

Table 5. *The relationship between sterical structure and 'non-specific' negative dromotropic effect of labetalol-isomers in guinea-pig isolated left atria*

Antagonist	n	– log molar concentration of antagonist required to reduce maximum driving frequency by 50%
RR-isomer	4	4.07(4.70–3.44)
SS-isomer	4	4.32(4.44–4.18)
RS-isomer	4	3.94(4.35–3.53)
SR-isomer	4	4.10(4.10–3.91)

Results are expressed as mean (and 95% confidence limits). R.T. Brittain, G.M. Drew & G.P. Levy[17]

Contrary to the α- and β-adrenoreceptor blocking potencies (see Table 4) the non-specific negative dromotropic effect is not dependent on the sterical structure, indicating that the effects are of a different nature.

Table 5a. DRUGS WITH COMBINED CARDIOVASCULAR ACTIONS

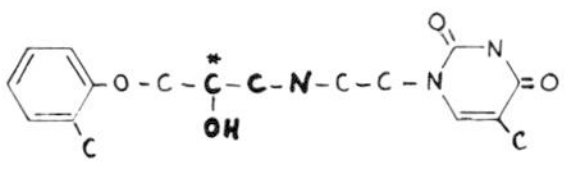

primidolol
β-blocker and α-blocker

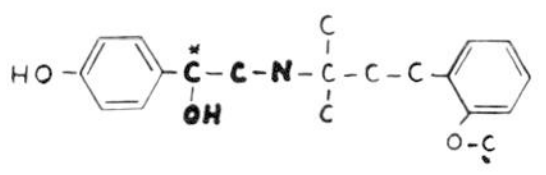

Draco(D 2343)
$β_2$-stim. and $α_1$-blocker

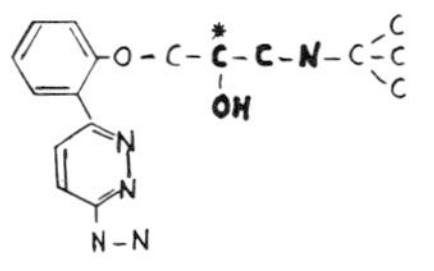

prizidilol
β-blocker and vasodilator

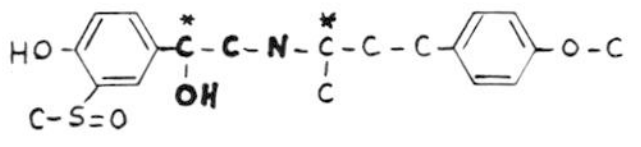

sulfinalol
β-blocker and vasodilator

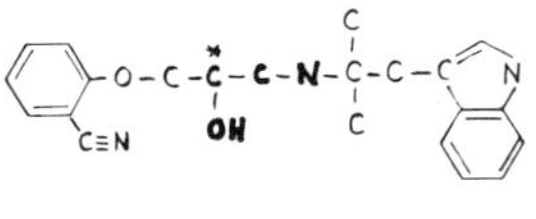

bucindolol
β-blocker, α-blocker and vasodilator

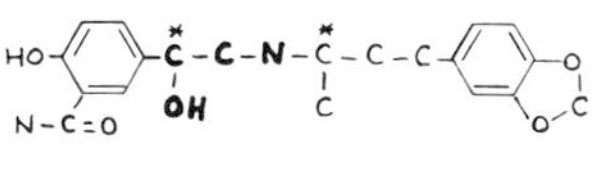

medroxalol (RMI 81968)
β-blocker, α-blocker and vasodilator

The information on stereochemical aspects of 'hybride compounds' in which different types of action, based on different mechanisms of action, are combined, can be extrapolated to the drugs with combined cardiovascular actions given in Table 5a. It has to be taken into account that the musculotropic vasodilatory action normally is not stereospecific, the eudismic ratio is close to 1. This has particular implications taken into account the stereoselectivity of the actions on α- and β-adrenergic receptors and the fact that the eudismic ratio for the α-adrenergic blocking action shows an inversion with regard to the other actions on adrenergic receptors.

One of the most striking examples of the separation of two components in the action in antipodes is the convulsive action in the (+)isomers of certain barbiturates such as DMBB 5-(1,3-dimethylbutyl)-5-ethylbarbiturate whereas the (-)isomers are anaesthetics and in fact antagonists (although probably not competitive) of the convulsant effects of the (+)isomers[18].

The foregoing exemplifies how different components in the action of 'one drug' may be distributed in various ways over the composing isomers. In some cases, like that of isoprenaline behaving as a partial agonist, the basic components in the action of the drug, the α-adrenergic and the α-adrenergic blocking action, are not even recognized in the racemic mixture.

Stereoselectivity as a tool in the analysis of mechanisms of action

Chemical manipulations resulting in a switch from agonistic to competitive antagonistic properties, namely introduction of alkyl and aralkyl substituents of increasing size in a particular position in the agonistic molecule, are not uncommon. This is possible, for instance, with histamine, leading to H_1-type antihistamines and with acetylcholine, resulting in atropine-like anticholinergics[19,27]. In such series of homologous compounds, one observes with an increase in the substituent a gradual decrease in both the apparent affinity and intrinsic activity of the compound. This initially leads to poorly active antagonists. With the introduction of the larger substituents, particularly aralkyl groups with one or two aromatic rings, however, the apparent affinity greatly increases such that higly potent antagonists are obtained (Table 6)[14,15,19,27]. An intriguing aspect of this switch from the cholinergic acetyl-β-methylcholine to anticholinergics is the stereochemistry. The center of asymmetry in acetyl-β-methylcholine heavily counts for the cholinergic action where the eudismic ratio is about 200 (Table 7). Once converted to an anticholinergic in the way indicated, this center of asymmetry hardly counts anymore (Table 7)[14,15,19]. This indeed indicates that in the cholinergic agent the choline moiety is essentially involved in the drug-receptor interaction, a high degree of complementarity to the receptor is required for that part of the molecule. In the anticholinergic agent the sterical configuration in the β-methylcholine moiety is irrelevant. The ring-bearing substituent, however, essentially contributes to the affinity (Table 6). This is confirmed by the fact that a center of asymmetry in that part of the molecule results in a relatively high degree of stereoselectivity again. There is a switch from a high stereoselectivity for the center of asymmetry in the choline moiety for the agonist to high stereoselectivity for the center of asymmetry in the ring-bearing moiety for the antagonist (Table 7)[14,15,19].

Introduction of both centers of asymmetry in one molecule gives an opportunity for further analysis. Some results are summarized in Fig. 4[14,15]. They confirm that differences in the configuration on the center of asymmetry in the ring substituted moiety clearly count. Differences in the center of asymmetry in the choline moiety, – of primary importance in cholinergics – hardly count in these anticholinergics. As shown schematically in Fig. 5[14,15,19] both in the case the α-adrenergic blocking aral-

Table 6. *The switch from a cholinergic to an anticholinergic action by molecular manipulation on acetyl-β-methylcholine.*

R–C(=O)–O–C(C)–C–N^{+}(C)$_3$			
R	i.a.	$pD_2 \pm P_{95}$	$pA_2 \pm P_{95}$
C–	1	6.3 ± 0.06	
C–C–	1	5.1 ± 0.14	
cyclo-(C–C–C)–	0.6 – 0.8	4.2 ± 0.4	
C–C–C–	0.1		3.8 ± 0.24
(C)$_2$C–	0		3.8 ± 0.23
C–C–C–C–C–	0		4.6 ± 0.26
C_6H_5–C(CH_2OH)(C)–	0		7.4 ± 0.20
$(C_6H_5)_2$C(OH)–	0		8.3 ± 0.13
C_6H_5(C_6H_{11})C(OH)–	0		8.9 ± 0.23

The ± figures give the P_{95} for the mean value for pD_2 and pA_2 obtained by testing of the compounds on the isolated gut of the rat.
Note that increase in the size of R results in a loss of intrinsic activity and strong decrease in apparent affinity (pA_2), resulting in poorly active anticholinergics. Introduction of the phenyl groups results in a strong increase in pA_2 such that highly potent anticholinergics are obtained. On basis of data from Ellenbroek et al.[14] and of refs. 15, 33, 35.

kyl derivatives of norepinephrine and the anticholinergic ring substituted derivatives of acetylcholine the hydrophobic aromatic rings do not bind to the relatively polar receptor sites on the receptor molecules, complementary to norepinephrine and the acetylcholines, but to accessory, auxiliary, binding areas located on the corresponding receptor molecules. These accessory hydrophobic binding areas on the receptor molecule, by their nature tend to form part of the interface of the receptor molecule and the lipid bilayer in which it is embedded. Binding of the antagonists under discussion will tend to stabilize the hydrophobic relationship between the receptor molecule and the lipid annulus around it and thus stabilize the solitary character of

Table 7. *Activity ratios of the stereoisomers of ester of β-methylcholine and choline tested on the isolated gut of the rat the high activity ratio – eudismic ratio – for the isomers of acetyl-β-methylcholine and the ratio close to 1 for the isomers of the benzylic acid ester of β-methylcholine. This indicates a loss of relevance of the center of asymmetry in the β-methylcholine moiety in this switch from cholinergic to anticholinergic action. The relatively high eudismic ratio for the center of asymmetry present in the anticholinergic phenylcyclohexyl-glycolic acid ester of choline indicates the relevance of the ring substituted group in the anticholinergic for the binding to its receptors.*
Based on data of Ellenbroek et al.[14] *and refs. 15, 28, 34, 38.*

$pD_2 \pm P_{95}$		config.	activity ratio	config.		$pD_2 \pm P_{95}$
7.0						7.0
6.8 ± 0.14		S_B	320	R_B		4.1 ± 0.23
$pA_2 \pm P_{95}$						$pA_2 \pm P_{95}$
8.0 ± 0.14		S_B	5/6	R_B		8.1 ± 0.10
8.6 ± 0.18						8.6 ± 0.18
9.6 ± 0.26		R_A	25	S_A		8.2 ± 0.14

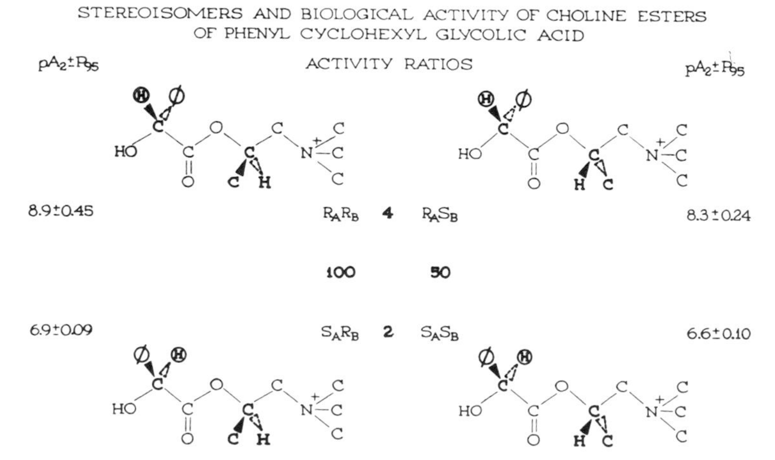

Figure 4. *The ratios for the anticholinergic action – eudismic ratios – of the four isomers of phenylcyclohexyl-glycolic-β-methylcholine tested on the jejunum of the rat (± figures give the P_{95} for the mean).*
Note: the ratios are small for the center of asymmetry in the β-methylcholine moiety and large for the center of asymmetry in the ring-substituted acyl moiety. On basis of data from Ellenbroek et al.[14] *and of refs. 15, 28, 33, 35.*

the receptor. Highly polar agonist molecules will tend to promote polar characteristics of the receptor protein and thus destabilize its relationship to the membrane lipids. This brings about a tendency for aggregation among the receptor molecules or among these molecules and other membrane proteins exposing suitable polar groups[21,22]. This concept fits in quite well with the receptor dynamics, 'mobile receptors', postulated in the various models for receptor-effector coupling[20–22].

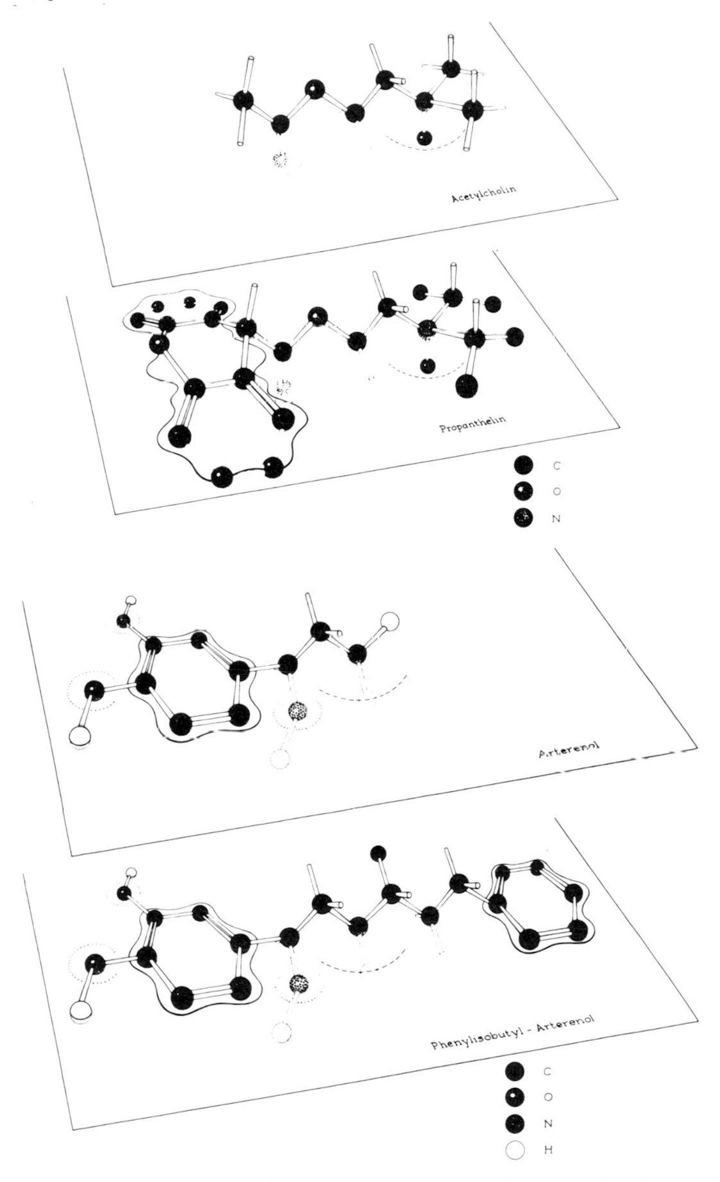

Figure 5. *Schematic representation of the relationship between the chemical structure of the agonists acetylcholine and noradrenaline and chemically related antagonists, the anticholinergic propantheline and the α-adrenergic blocking agent phenylisobutylnoradrenaline and their binding to the respective receptor sites.*
Note: the hydrophobic aromatic substituents in the antagonistic molecules bind to accessory areas on the receptor molecule, not a part of the polar receptor sites, complementary to the polar agonistic molecules[15,16,32].

Sterical structure and selectivity in action

Besides stereoselectivity in action that is a high eudismic ratio for optical isomers with regard to a particular effect, there is the selectivity with regard to different types of action. In various types of competitively blocking agents such as anticholinergics, α-adrenergic blockers, H_1-antihistamines, etc. the moiety bearing the unsaturated rings essentially contributes to the affinity, it is a determinant factor. The accessory (auxiliary) binding areas on these various receptors are predominantly hydrophobic in nature[15,16,19]. Many compounds are found with multiple actions, for instance, a simultaneous H_1-antihistaminic and anticholinergic action. This will especially hold true for compounds in which the hydrophobic moiety is rather flexible such that it easily adapts to the stereochemistry of the accessory binding areas on both the H_1- and the Ach receptor. An example is diphenhydramine in which both aromatic rings and the carbon chain allow a high degree of freedom for the conformation of the molecule (Table 8)[23]. Introduction of alkylsubstituents in position 2 or 4 in one of the rings results in a remarkable selectivity. Introduction in the 2 position of alkyl substituents of increasing size results in an increase in the anticholinergic activity and a decrease in the antihistaminic activity. No doubt, the tertiary butyl group in this position strongly reduces the free rotation of the rings with as a consequence predominance of a conformation which apparently fits better to the accessory receptor area of the cholinergic receptor than to that of the H_1-histamine receptor. Introduction of an alkyl substituent in the 4 position of one of the rings results, particularly as long as small groups are involved, in an increase in the antihistaminic activity and a decrease in the anticholinergic activity. This parasubstitution possibly brings about a certain rigidity by hyperconjugation resulting in more planarity in the molecule. The substitutions imply introduction of a center of asymmetry.

Table 8. *The increase of the specificity in action of diphenhydramine derivatives as a result of chemical manipulation leading to restriction in the degree of conformational freedom. After data of Harms et al.*[23] *and refs. 32-34.*

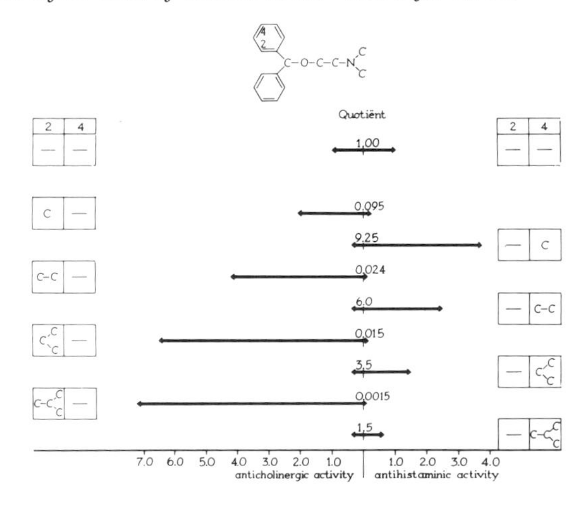

Introduction of alkyl substituents of increasing size in the 2-position forcing both rings out of one plane results in an increase in the anticholinergic and decrease in the antihistaminic activity. Introduction of small alkyl substituents in the 4-position may by hyperconjugation tend to bring this ring in one plane with the rest of the molecule resulting in an increase in the antihistaminic and a decrease in the anticholinergic activity.

The activity ratio in pairs of stereoisomers – eudismic ratios

A high biological activity, that is a high affinity, will require a high degree of complementarity between the molecules of the bioactive agent and the specific receptor sites involved in the action. This implies a strong interaction, a close approach, between the complementary groups. As a consequence for highly active compounds with a center of asymmetry, located in a part of the molecule essentially involved in the interaction with the receptors, a high eudismic ratio is expected for the stereoisomers. Small eudismic ratios are expected for molecules with a low affinity and thus a poor complementarity to the receptor sites. On basis of the foregoing, one expects that for diphenhydramine with a tertiary butyl group in 2 position, the eudismic ratio will be high for the anticholinergic activity and low for the antihistaminic activity while on the other hand for the compounds with a methyl group in the 4 position, the ratio will be large for the antihistaminic and low for the anticholinergic activity. This indeed is the case (Fig. 6)[24]. For compounds with a center

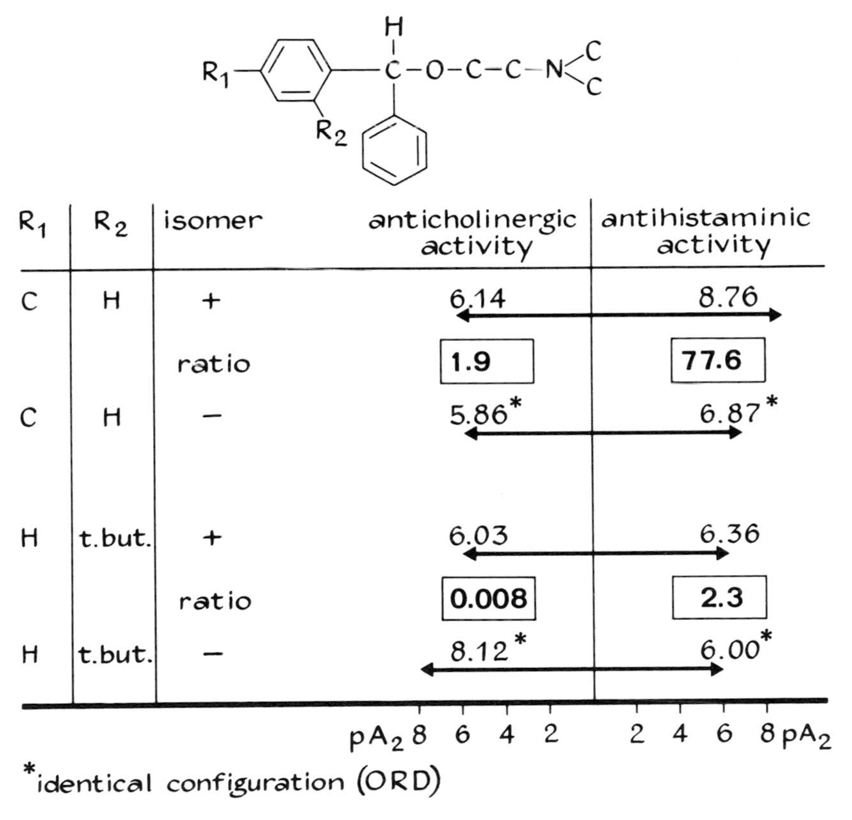

Figure 6. *Ratios for the activities of optical isomers – eudismic ratios – for some of the derivatives of diphenhydramine presented in table 8.*
Note: In the R_1*-methylsubstituted derivative the eudismic ratio is high for the relatively strong antihistaminic activity and low for the relatively poor anticholinergic activity. In the* R_2*-tertiary butyl substituted derivative, the eudismic ratio is high for the relatively strong anticholinergic activity and low for the relatively poor antihistaminic activity. On basis of data of Rekker et al.*[24].

of asymmetry in a part of the bioactive molecule not closely involved in the interaction with the receptor site too, small ratios are to be expected. Accordingly for the type of anticholinergics as described in Table 6 and 7 and Fig. 4, high eudismic ratios are found for compounds with the center of asymmetry in the ring-bearing moiety and well, ratios which are high for compounds with high pA_2-values and low for compounds with low pA_2-values. This phenomenon is manifested clearly if the eudismic ratios are plotted against the pA_2-value for the eutomer (Fig. 7)[15,21]. This type of relationship appears to be common for many types of bioactive agents[9,10] and is often called Pfeiffer's rule since he already in 1956 reported on this relationship for a variety of drugs used in medical practice[25].

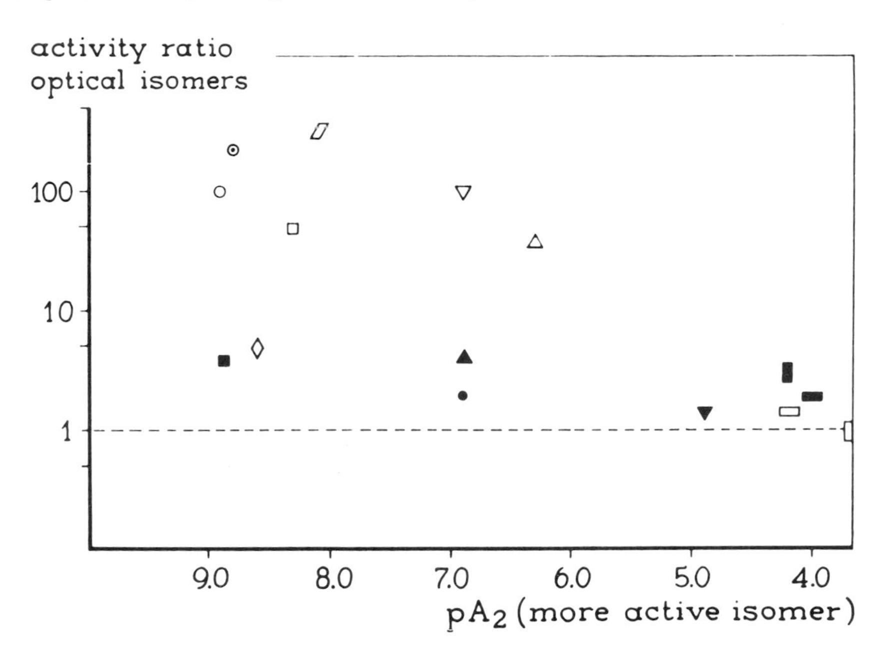

Figure 7. *The relationship between the activity ratios – eudismic ratios – for pairs of stereoisomers and the pA_2-values of the most active isomers – the eutomers – of various anticholinergic compounds tested on the isolated gut of the rat. The open symbols concern eudismic ratios related to a center of asymmetry in the ring bearing moiety, the closed symbols to the ratios related to a center of asymmetry in the aminoalcohol moiety.*
Note: the eudismic ratio for the compounds with a center of asymmetry in the aminoalcohol moiety – a non critical moiety – is not dependent on the pA_2-value of the eutomer. Such a dependence clearly exists, however, for the compounds with a center of asymmetry in the ring bearing moiety – a critical moiety: An exception is the phenyl-thienyl-glycolic acid ester of choline (◇). Taken into account the isosteric, isofunctional, character of the phenyl and the thiophen group, the center of asymmetry in this compound is of chemical but not of biological relevance. For detailed information on the various compounds see the references[14,15,28,33,34,35].

As expected on basis of the analysis of structure-action relationship for the anticholinergics, discussed before, independent of the pA_2-values, low eudismic ratios are found for the compounds with the center of asymmetry located in the choline moiety, a part of the molecule not closely involved in the interaction with the receptor. There is a remarkable exception in Fig. 7, namely the phenyl-thiophenglycolic acid ester. For this compound the eudismic ratio is close to one, not withstanding the high pA_2-value. Biologically the phenyl group and the thiopen group are isosteric, or even better, isofunctional. This means that substitution of the one by the other has little or no influence on the biological activity. Thus exceptions to Pfeiffer's rule are found for compounds with a center of asymmetry in a part of the molecule not essentially involved in the interaction with the receptors and isosterism, biological isofunctionality, of two of the groups on the center of asymmetry. One further has to take into account that the eudismic ratio and thus the relationship described may be obscured by incomplete separation of the isomers. This particularly counts for the presence of certain amounts of the eutomer in the distomer. The reverse situation is less disturbing.

Nevertheless the relationship outlined is observed for a wide variety of bioactive agents, such as antihistamines, β-adrenergic and β-adrenergic blocking agents, narcotic analgetics, various types of bioactive polypeptides and auxin type herbicides (Fig. 8)[10].

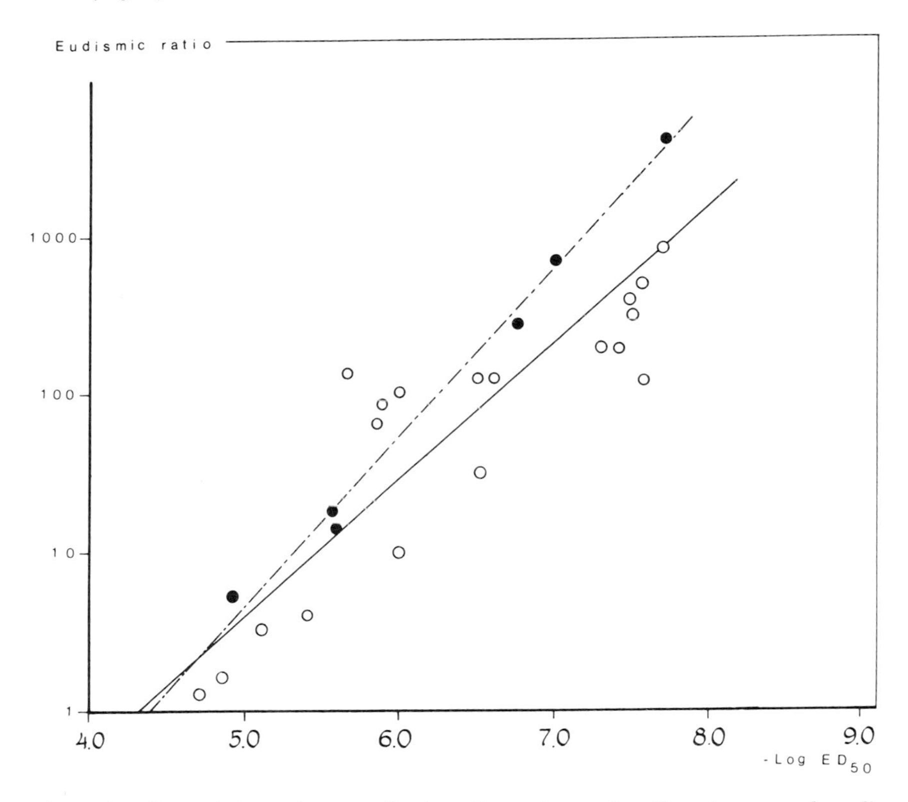

Figure 8. *The activity ratios – eudismic ratios – for auxine-like phenoxycarboxylic acids (o) and for β-naphthoxycarboxylic acids (•) as a function of the negative logarithm of the ED_{50} of the most active isomer – the eutomer – tested as inhibitors of the growth of flax-root*[10]. *For detailed information on the various compounds see the refs. 36, 37.*

The phenomenon is not restricted to interactions of bioactive agents with receptors but also holds true for enzymes. Particularly for enzyme inhibitors where binding constants and not turn-over rates count. An example is given in Fig. 9[6], where eudismic ratios for the rate of inactivation of acetylcholine-esterase and butyrylcholine-esterase (pseudocholine-esterase) by stereoselective organic phosphates are plotted against the rate of inactivation of the enzymes by the eutomer. Remarkably, there is a clearcut dependence only for the true acetylcholine-esterase. On the whole, for pseudocholine-esterase the rate of inactivation by the various compounds is lower than for the true acetylcholine-esterase. The reason for this difference may be that on the active site of acetylesterase, three of the groups on the chiral phosphor atom in the organic phosphates participate in the interaction, while in the case of the pseudocholine-esterase only two of these groups, among which the thioalkylamino group, are essentially involved.

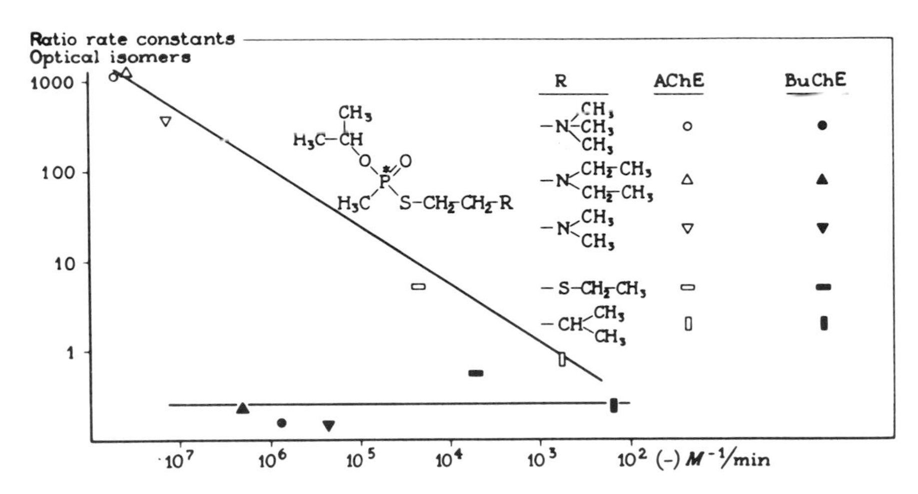

Figure 9. *The relationship between the ratios of the activities of pairs of sterical isomers - eudismic ratios - of a series of organic phosphates with a center of asymmetry on the phosphor atom and the activity of the more active isomer - the eutomer - in each of the pairs. The compounds were tested as inhibitors of acetylcholinesterase (true cholinesterase; open symbols) and on butyryl cholinesterase (pseudocholinesterase; closed symbols)*[5,6].
Note: the dependence of the eudismic ratio on the activity of the eutomers for acetylcholinesterase and the absence of such a dependence for pseudocholinesterase[28].

Elucidative reviews on the aspects of stereoselectivity discussed in this section are found in the literature[10,10a,26].

Position-isomers and bulk-tolerance

Structure-action relationships for cholinergic and anticholinergic derivatives of choline and β-methylcholine, discussed in the foregoing sections make probable that the fit of the choline moiety and thus the onium group in the anticholinergics to the corresponding site on the receptor is rather loose, although it contributes to the binding. In contrast to the cholinergic agents where on the onium group only methyl groups or maximally introduction of one ethyl group is well tolerated, in the corresponding anticholinergic agents, substitution even with larger alkyl groups is well tolerated. The onium group even may be built in in large structures, such as the tropinol group.

Although the distance between the onium group and its complement on the receptor site in anticholinergics apparently is relatively large, introduction of sufficiently bulky alkyl groups will re-establish close contact between the substituted onium group and the receptor site. It is feasible that the onium group will approach the receptor site as close as possible. The carbon skeleton of the tropinol moiety therefore will tend to be away from the receptor. The space left open between the onium group and its corresponding group on the receptor site is determinant then for the spatial tolerance with regard to alkyl substitution on the onium group. Fig. 10 gives experimental evidence[21,28,38]. Two series of 'position' isomers of the anticholinergic atropine are presented. Introduction of a methyl group is well tolerated in both positions. The activity is practically equal to that of atropine. Introduction of alkyl groups of increasing size in position R_1, directly results in a decrease of activity. In position R_2, an ethyl group and even an isopropyl group are well tolerated. As a consequence, for the isopropyl derivates in both series of position isomers, a large 'eudismic' ratio is found. Substitution of a propyl group also in position R_2 results in a loss of activity.

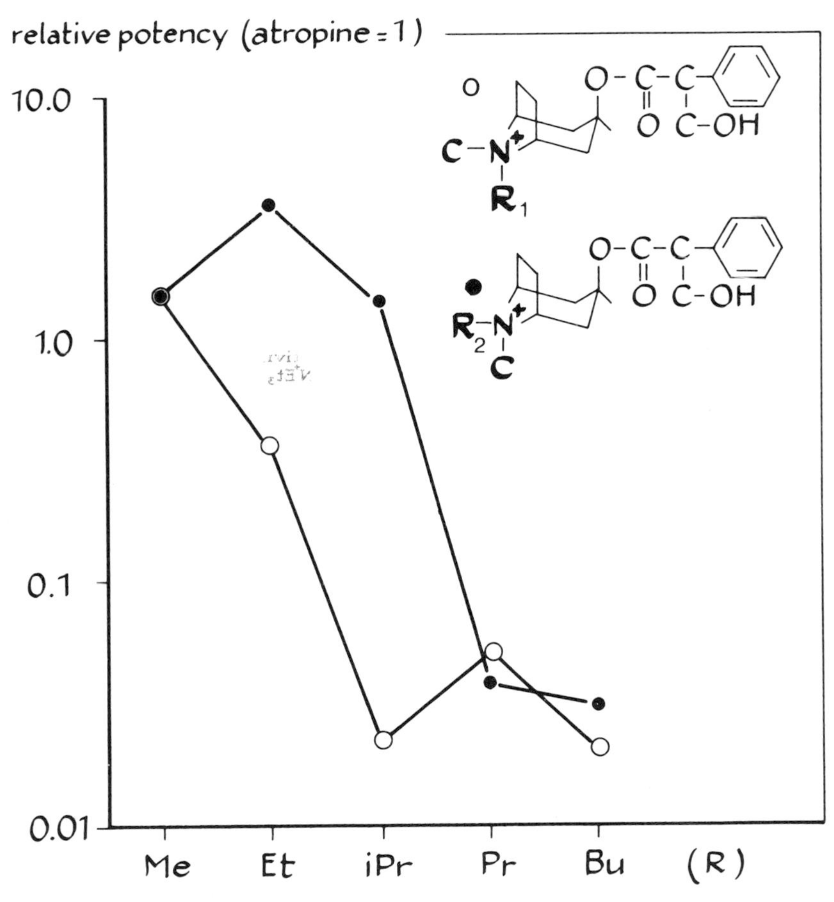

Figure 10. *The activities of N-alkylsubstituted derivatives of atropine, constituting two series of geometrical isomers (position isomers). The compounds are tested against acetylcholine on the isolated gut of the guinea pig.*
Note: the large ratio, ± 100, for the activities of the isopropylsubstituted derivatives. There is a clear difference in tolerance for alkyl substitution in the two series of position isomers. On basis of data of Wick[31]. See also refs. 21, 38.

Not only for centers of asymmetry it counts that the impact thereof on activity is dependent on the location in moieties of the molecule critically involved in the receptor interaction. This also holds true for the impact of chemical substitutions, for instance, ethyl substitution in the trimethylammonium group of cholinergic and anticholinergic agents. As shown in Fig. 11[15,30,31,35] this chemical manipulation hardly changes the activity for the diphenylacetyl type trimethylammonium anticholinergics. It heavily counts, however, for the cholinergic agents, where it brings about a shift to poorly active anticholinergics, this with the exception for the mono-ethylated derivatives. It is assumed that one ethyl group is tolerated because it can be in a position away from the receptor site.

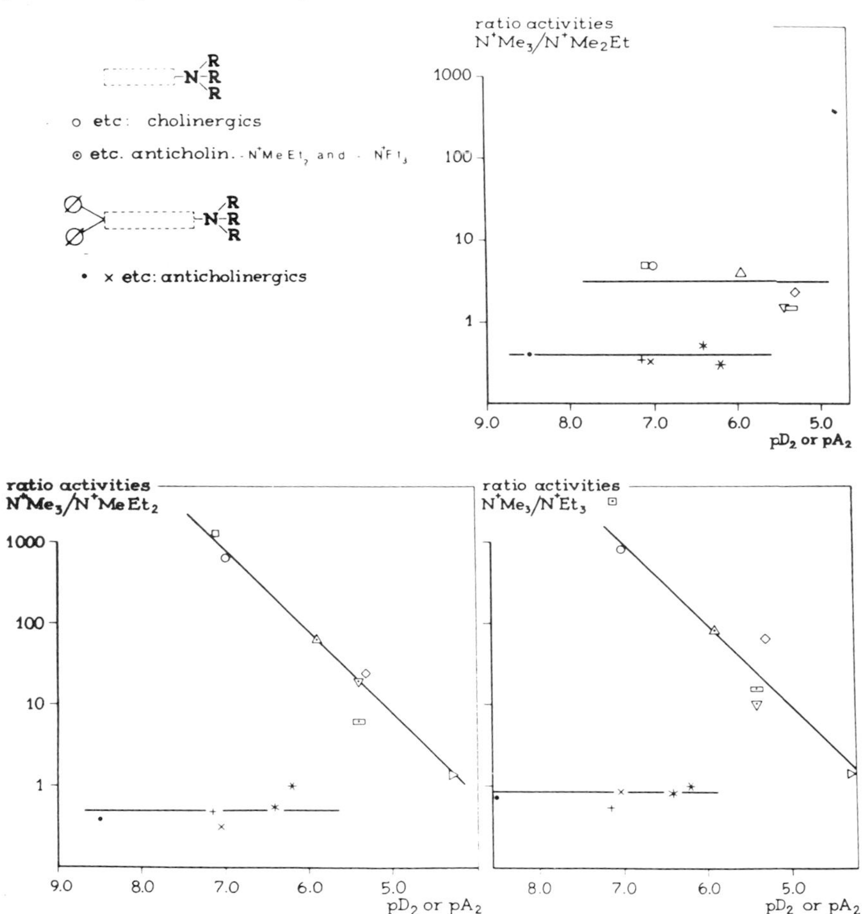

Figure 11. *The relationship between the ratios of activities of pairs of compounds – differing only in the substitution on the onium group – and the activity of the corresponding -NMe_3-derivative plotted on the abcis. The open symbols represent compounds structurally closely related to acetylcholine, the closed symbols represent anticholinergic agents with one or two phenyl rings in the moiety opposite to the onium function.*
Fig. a) concerns the ratios for the -N^+Me_3 and -N^+Me_2Et derivatives
Fig. b) concerns the ratios for the -N^+Me_3 and -N^+MeEt_2 derivatives
Fig. c) concerns the ratios for the -N^+Me_3 and -N^+Et_3 derivatives
The cholinergic and anticholinergic activities of the various compounds are tested on the isolated gut of the rat. Part of the compounds structurally closely related to acetylcholine act as cholinergic (open symbols), part of these compounds act as weak anticholinergics (dotted open symbols). The ring substituted derivatives act as, partly potent, anticholinergics (closed symbols). For details on the various compounds see original publications[15,28,30,35].

Note: in the anticholinergic agents (closed symbols) the switch from -N^+Me_3-derivatives to the respective ethylated derivatives hardly changes the activity, the ratios are close to one. For the compounds structurally related to acetylcholine, however, only substitution of one ethyl group is tolerated without a great change in the activity (pD_2 or pA_2 values). Further ethylation results in large ratios for the activities of the pairs of compounds, dependent on the activity, pD_2 or pA_2 of the -N^+Me_3-derivative. This indicates that the onium group is in a more critical position on the receptors. Notice the analogy in the phenomena observed for pairs of stereoisomers (Fig. 7, 8 and 9) and those represented in this figure.

Here too, Pfeiffer's rule shows up, which is not unexpected since strictly taken chemical manipulations of the type just discussed in fact too are changes in the 'sterical properties' of the molecules, be it not in the sense of sterical antipodes.

References

1. L.H. Easson and E. Stedman, *Biochem. J. 27,* 1257 (1933).
2. P.S. Farmer and E.J. Ariëns, *TIPS 3,* 362 (1982).
3. M. Robertson, *N. Sci. 94,* 696 (1982).
4. W. Eberlein, *in:* Arzneimittelentwicklung: Grundlagen-Strategien-Perspektiven, p. 2 (E. Kutter, ed.), Georg Thieme Verlag, Stuttgart, 1978.
5. H.L. Boter, Thesis, University of Leiden, 1970.
6. A.J.J. Ooms and H.L. Boter, *Biochem. Pharmacol. 14,* 1839 (1965).
7. H.P. Benschop, F. Berends and L.P.A. de Jong, *Fundam. Appl. Toxicol. 1,* 177 (1981).
8. R. Wegler and L. Eue, *in:* Chemie der Pflanzenschutz- und Schädlingsbekämpfungsmittel, Vol. 2, p. 278 (R. Wegler ed.), Springer-Verlag, Berlin, 1970.
9. L.C. Luckwill and D. Woodcock, *in:* The chemistry and mode of action of plant growth substances, p. 195 (R.L. Wain and F. Wightman eds.), Butterworths Scientific Publications, London, 1956.
10. P.A. Lehmann F., J.F. Rodrigues de Miranda and E.J. Ariëns, *in:* Progress Drug Research, Vol. 20, p. 101 (E. Jucker ed.), Birkhäuser Verlag, Basel and Stuttgart, 1976.

10a. J.F. Rodrigues de Miranda, P.A. Lehmann F. and E.J. Ariëns, *in:* Progress Drug Research, Vol. 20 p. 126 (E. Jucker ed.), Birkhauser Verlag, Basel and Stuttgart, 1976.

11. E.J. Ariëns, *Naunyn-Schmiedebergs Arch. Pharmak. exp. Path. 257,* 118 (1967).
12. E.J. Ariëns, *Procs. First Int. Pharmacol. Meeting 7,* 247, Pergamon Press, Oxford, 1963.
13. P.N. Patil, D.D. Miller and U. Trendelenburg, *Pharmacol. Revs. 26,* 323 (1975).
14. B.W.J. Ellenbroek, R.J.F. Nivard, J.M. van Rossum and E.J. Ariëns, *J. Pharm. Pharmacol. 17,* 393 (1965).
15. E.J. Ariëns and A.M. Simonis, *Ann. NY Acad. Sci. 144,* 842 (1967).
16. E.J. Ariëns, *Ann. NY Acad. Sci. 139,* 606 (1967).
17. R.T. Brittain, G.M. Drew and G.P. Levy, *Br. J. Pharmac. 77,* 105 (1982).
18. I.K. Ho and R. Adron Harris, *Ann. Rev. Pharmacol. Toxicol. 21,* 83 (1981).
19. E.J. Ariëns, *in:* Advances in Drug Research, Vol. III, p. 235 (N.J. Harper and A.B. Simmonds eds.), Academic Press, London, 1966.
20. P. Cuatrecasas, M.D. Hellenberg, K.J. Chang and V. Bennett, *Rec. Prog. Horm. Res. 31,* 37 (1975).
21. E.J. Ariëns, A.J. Beld, J.F. Rodrigues de Miranda and A.M. Simonis, *in:* The Receptors, Vol. I: General Principles and Procedures, p. 33 (R.D. O'Brien ed.), Plenum Press, New York-London, 1979.
22. E.J. Ariëns, *TIPS 1,* 11 (1979).
23. A.F. Harms and W.Th. Nauta, *J. Med. Pharm. Chem. 2,* 57 (1960).
24. R.F. Rekker, H. Timmerman, A.F. Harms and W.Th. Nauta, *Arzneim.-Forsch. 21,* 688 (1971).
25. C.C. Pfeiffer, *Science 124,* 29 (1956).
26. P.A. Lehmann F., *in:* Receptors and Recognition, Series A, Vol. 5, (P. Cuatrecasas and M.F. Greaves eds.), Chapman & Hall, London, 1979.
27. E.J. Ariëns and A.M. Simonis, *Arch. Int. Pharmacodyn. Ther. 127,* 479 (1960).

28. E.J. Ariëns, *in:* Drug Design, Vol. 1, p. 162 (E.J. Ariëns ed.), Academic Press, New York-London, 1971.
29. E.J. Ariëns, A.M. Simonis and J.M. van Rossum, *in:* Molecular Pharmacology, Vol. 1, p. 242 (E.J. Ariëns ed.), Academic Press, New York, 1964.
30. A.M. Simonis, E.J. Ariëns and J.F. Rodrigues de Miranda, *Acta Physiol. Pharmacol. Neerl. 12,* 300 (1963).
31. H. Wick, unpublished data. Research Institute C.H. Boehringer Sohn, Ingelheim am Rhein, Germany, 1972.
32. E.J. Ariëns, *Chem. Weekbl. 56,* 301 (1960).
33. E.J. Ariëns, *Arzneim.-Forsch. 16,* 1376 (1966).
34. E.J. Ariëns and A.M. Simonis, *Procs. IIIrd Int. Pharmacol. Meeting 7,* 271, Pergamon Press, Oxford-New York, 1968.
35. E.J. Ariëns, *Actual. Pharmacol. 22,* 261 (1969).
36. A. Jönsson, *in:* Handbuch der Pflanzenphysiologie, Vol. 14, p. 999, (W. Ruhland ed.), Springer-Verlag, Heidelberg, 1961.
37. B. Äberg, *in:* Plant Growth Regulation, International Conference, 4th ed., p. 219, Yonkers, New York, 1961.
38. E.J. Ariëns, *in:* Schriftenreihe der Bundesapothekerkammer zur wissenschaftlichen Fortbildung, Vol. IV, p. 77, Meran, 1976.

Stereoselectivity in Drug and Xenobiotic Metabolism

N.P.E. Vermeulen and D.D. Breimer

Abstract

Presently it is well recognized that stereochemical factors play a significant role in the metabolism of drugs and other xenobiotics. The reason for this is that binding of a racemic modification of a chiral molecule to an optically active macromolecule (e.g. a drug metabolizing enzyme) results in the formation of two diastereomeric complexes with different physical and chemical properties. As a result of this the enantiomers of chiral substrates may be metabolized at different rates (i.e. substrate stereoselectivity); alternatively, the creation of a new asymmetric center during the metabolic process may result in the formation of different stereoisomers at different rates (i.e. product stereoselectivity).

Stereospecific metabolism, which by definition means completely stereoselective, is only rarely observed in drug metabolism. However, reports on stereoselectivity metabolism of drugs and other xenobiotics are numerous. The various drug metabolizing enzymes such as mixed function oxidases, epoxide hydrolases as well as glutathione transferases all have been demonstrated to exhibit substrate- and productstereoselective effects. Both phenomena may give rise to differences in pharmacological and toxicogical properties of the stereoisomers.

In this communication a number of examples will be discussed to illustrate the particular aspects of stereoselective drug metabolism. The examples will refer to drugs used in daily practice as well as to toxic xenobiotics.

Introduction

It is presently well recognized that stereochemical factors play a significant role in the disposition of body-foreign substances in living organisms. Desired or undesired effects of drugs or other xenobiotics may be influenced by stereoselective effects of their uptake, distribution, metabolism as well as excretion. The underlying reason for this is in principle the fact that binding of a racemic modification of a chiral molecule to an optically active macromolecule (e.g. a metabolizing enzyme) results in the formation of two diastereomeric complexes with different chemical and physical properties, such as transition state, stability of the complex and others. As a result, the rate or the nature of a metabolic reaction may become quite different for different stereoisomers.

Some years ago Jenner and Testa[1] further defined stereoselectivity in relation to drug metabolism in terms of substrate-stereoselectivity and product-stereoselectivity. The first term should be used in a situation where two enantiomers (optical isomers) of an asymmetric (chiral) substrate are metabolized at different rates, but without the introduction of a new asymmetric center. When during metabolism an (or another) asymmetric center is created and when the resulting enantiomers (or diastereoisomers) are formed at different rates the term product-stereoselectivity is used. The term stereospecific metabolism is by definition used for completely stereoselective metabolism and as such only very rarely observed in drug metabolism.

In this chapter a number of examples of stereoselective metabolism of drugs and other xenobiotics is discussed with the prime objective to illustrate the occurrence of stereoselective metabolism as well the importance in pharmacology and toxicology. Since the number of published examples of stereoselective metabolism is almost countless, attention is especially focussed on the stereoselective action of a number of important enzymes involved in the metabolism of drugs and other xenobiotics, rather than on examples of single compounds. Only for hexobarbital more detailed information, based on own research, will be given.

Importance of stereoselective metabolism

From a quantitative point of view stereoselective metabolism of drugs and other xenobiotics is an extremely important phenomenon. Extensive and elaborate reviews on this subject have been published in recent years[1-4].

Stereoselective metabolism of drugs may be important from a pharmacological or therapeutic point of view, particularly when enantiomers differ considerably in pharmacological properties, either qualitative or quantitative. For example, the biotransformation of the isomer which is responsible for the desired pharmacological effect may be influenced by enzyme inhibiting or otherwise pharmacodynamic effects, such as observed in the metabolism of amphetamine[5], propoxyphene[6] or propranolol[7]. On the other hand isomers may be eliminated at very different rates. The latter is for example the case for the hypnotic agent hexobarbital. In man the elimination half-life of the more active (+)-isomer is about three times longer than of the less active (-)-isomer (Fig. 1; [8]). This was due to a difference in hepatic metabolic clearance and not to differences in volumes of distribution or plasma protein binding between the enantiomers. In rats the reversed situation was found with regard to the rate of elimination of hexobarbital enantiomers. Such species dependent differences in stereoselective metabolism have also been observed for pentobarbital[9], amphetamine[1,3] and warfarin[4]. Although it is difficult to generalize it is likely that the multiplicity, in combination with substrate-selectivity of metabolizing enzymes are responsible for the observed differences in stereoselectivity between animal species[10,11].

Product-stereoselective metabolism may also be important from a toxicological point of view. This was for example illustrated by experiments in our institute on mutagenic behaviour of *cis-* and *trans-* 1,2-dichlorocyclohexane with and without metabolic activation[12]. Using the Ames-test with Salmonella typhimurium cells it was shown that, in contrast to the *trans*-isomer, the *cis*-isomer demonstrated mutagenic activity that was significantly increased upon addition of 9000 *g* and 100,000 *g* supernatant of a rat liver homogenate (Fig. 2; [12]). The explanation of this

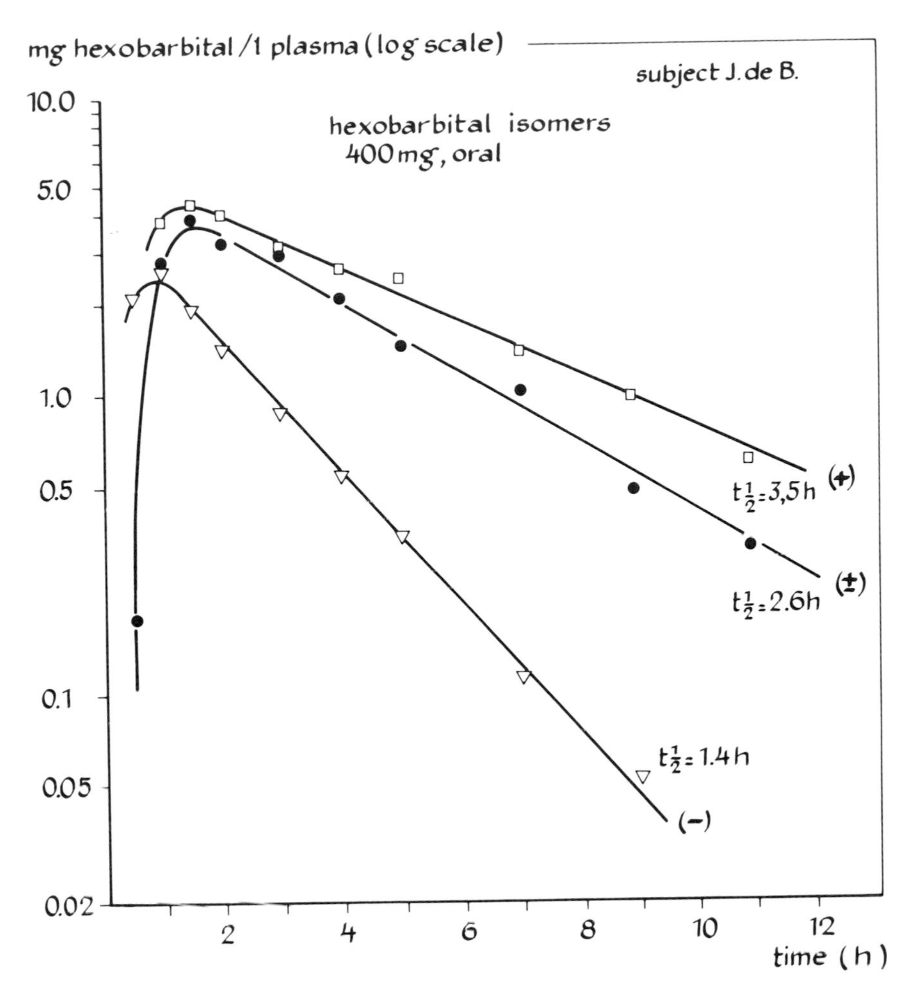

Figure 1. *Blood-concentration time curves of (+)-and (−)-hexobarbital in the same volunteer; 400 mg hexobarbital-enantiomers ((+) and (−)) or racemic hexobarbital (±) were given orally on separate occasions (taken from Breimer et al.*[8]*)*

stereoselective effect can be given in terms of a stereospecific action of glutathione transferases present in the liver homogenate fractions and catalyzing the substitution of the vicinal chlorine atoms according to a Sn_2-substitution mechanism. Only in the *cis*-1,2-dichlorocyclohexane this conjugation reaction can lead to the formation of a reactive thiiranium-ion, which is suspected to be responsible for the mutagenic activity (Fig. 3; [12]).

Although generally biotransformation enzymes play an important role in the detoxification of drugs and other xenobiotics, it should be realized that particular structural properties of a compound in combination with a stereoselective action of the enzymes involved sometimes give rise to bioactivation (toxification) of xenobiotics[13–15]. Both aspects will be illustrated further in this chapter.

Stereoselectivity in mixed function oxidases

One of the most important drug and xenobiotic metabolizing enzyme systems in mammalians and various non-mammalians is the mixed-function oxidase system

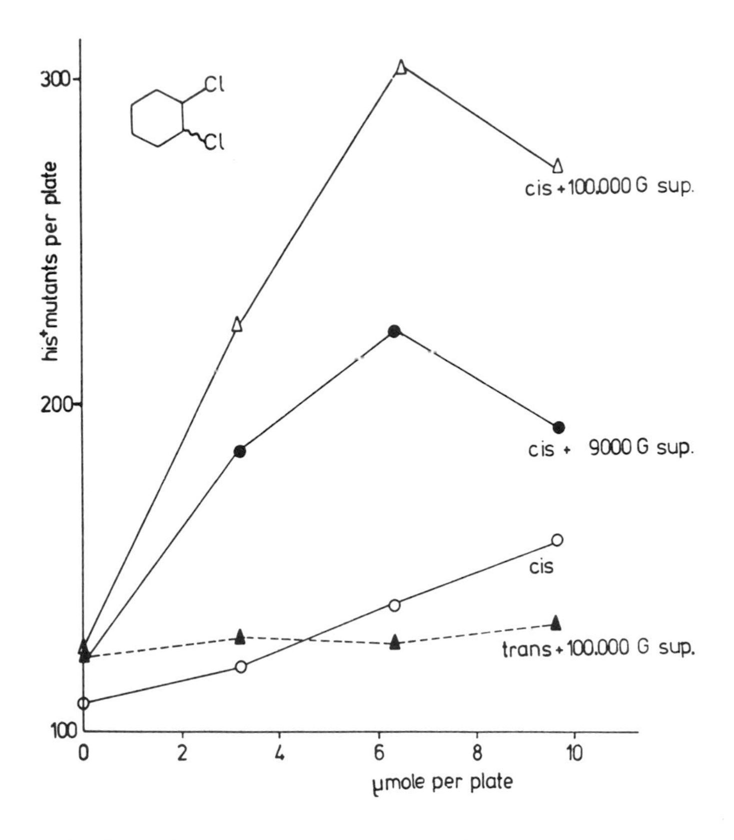

Figure 2. *Number of reverse mutations in Salmonella typhimurium TA100, induced by* cis- *and* trans-*1,2-dichlorocyclohexanes with and without the addition of homogenated rat liver fractions. For the* trans-*isomer only the experiment with the addition of 100,000 g supernatant and GSH is shown (taken from Van Bladeren et al.*[12]*)*

(MFO- or cytochrome P-450 system). It is ubiquitously distributed among various organs of mammalians, e.g. liver, lung, skin etc, and it consists of several cytochrome P-450 containing forms with broad and overlapping substrate-selectivities[14,16]. The MFO-system catalyzes a wide variety of metabolic reactions, such as epoxidations, hydroxylations, dealkylations, deaminations as well as reduction reactions[17]. Despite the apparent broadly ranging substrate- and product-selectivity of this enzyme system, numerous examples have been reported illustrating stereoselectivity in the metabolism of various drugs and other xenobiotics[1,3,18].

Among the best documented examples illustrating stereoselective action of cytochrome P-450 is the one recently described by Jerina *et al.*[19]. They investigated extensively the stereoselective metabolism of polycyclic aromatic hydrocarbons by the intact MFO-system, as well as by single purified forms of rat liver cytochrome P-450. One of these forms, the so-called cytochrome P-450c catalyzed highly stereoselectively the epoxidation of benzo(a)pyrene to three regioisomeric benzo(a)-pyrene-oxides with the absolute configuration 4S, 5R, 7R, 8S and 9S,10R (Fig. 4; [20]). In contrast to the 4S,5R- and 9S,10R-oxides, benzo(a)pyrene-7R,8S-oxide is a precursor in the bioactivation process of benzo(a)pyrene. After stereoselective

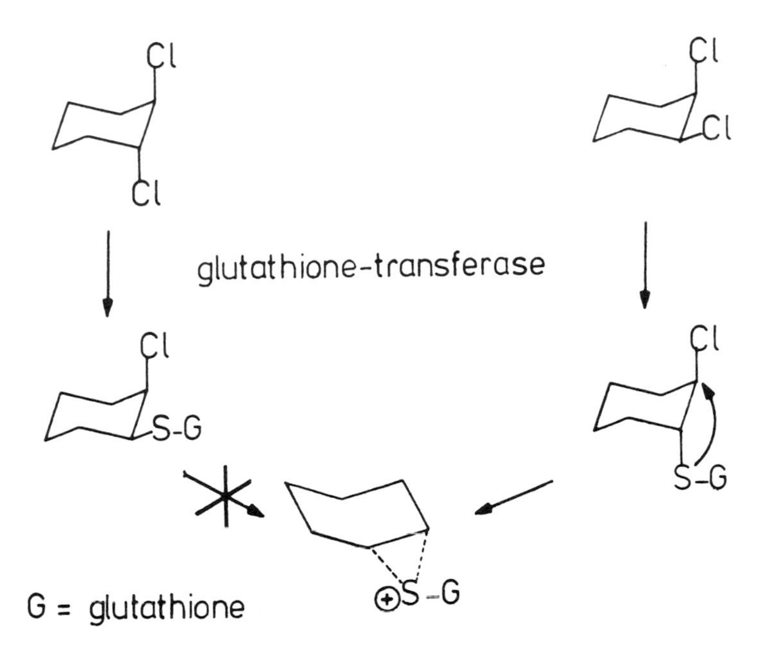

Figure 3. Conjugation of cis- *and* trans-*1,2-dichlorocyclohexane to glutathione. Only with the* cis-*isomer the formation of the reactive thiiranium-ion is possible (taken from Van Bladeren* et al.[12]*)*

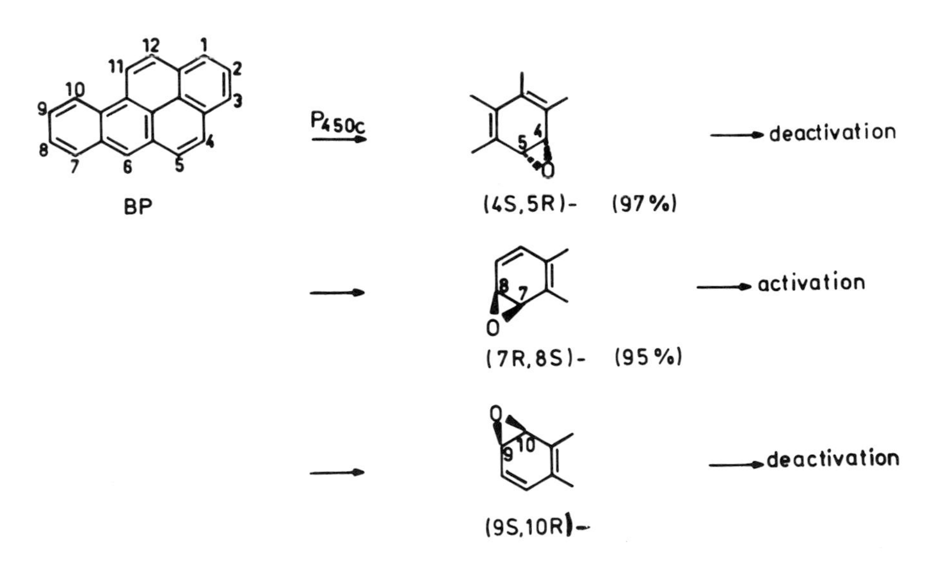

Figure 4. *Stereoselective formation of three regioisomeric arene oxides of benzo(a)-pyrene (BP) by purified cytochrome P-450c of rat liver (P-450c). The formation of the arene oxides proceeds highly stereoselective. Absolute configurations and enaniomeric purities are indicated in brackets (derived from Jerina* et al.[19]*).*

hydrolysis of this arene-oxide (called: proximate carcinogen) to (–)-7R, 8R-dihydro-diol-benzo(a)pyrene by epoxide hydrolase, cytochrome P-450c converts this diol to benzo(a)pyrene-7R,8S-dihydrodiol-9S,10R-epoxide, in which the benzylic hydroxyl group and epoxide oxygen are *trans,* in over 85% diastereomeric excess (Fig. 5). The latter diol-epoxide (called: ultimate carcinogen) was demonstrated to be a considerably more potent initiator of tumors than its stereoisomers. The major structural features rendering in particularly the highly unstable diol-epoxides of benzo(a)pyrene mutagenic and carcinogenic appear to be a combination of the diol-epoxide functionality in a trans-configuration (Fig. 5) bordering the 'bay-region' of benzo(a)pyrene, thus rendering the C_{10}-atom to react as a carbonium ion towards nucleophilic cellular components, notably the N_2-exocyclic amino-group of DNA-guanine[21,22].

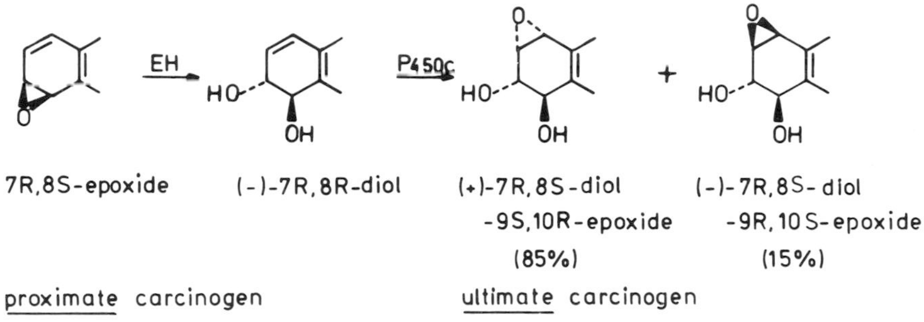

EH P450c +
7R,8S-epoxide (-)-7R,8R-diol (+)-7R,8S-diol -9S,10R-epoxide (85%) (-)-7R,8S-diol -9R,10S-epoxide (15%)
proximate carcinogen ultimate carcinogen

Figure 5. *Stereoselective formation of 7,8-dihydrodiol-9,10-epoxides from one benzo(a)pyrene-7,8-oxide by epoxide hydrolase (EH) and purified rat liver cytochrome P-450c (P-450c). Absolute configurations and diastereomeric percentages are indicated in brackets (derived from Jerina* et al.[19]*).*

By studying the stereospecificity of various polycyclic aromatic hydrocarbons in detail, Jerina and coworkers were able to propose a computer-simulated model for the steric requirements of the catalytic site of the cytochrome P-450c involved (Fig. 6; [19]). This relatively simple model could predict for benzo(a)pyrene: a) – the regioisomeric epoxidation, b) – the fact that the arene-oxides formed should have the R-absolute configuration at the benzylic centers and the S-absolute configuration at the allylic centers. In addition the model could predict, realizing that epoxide hydrolase converts the configuration of the allylic center in forming *trans*-dihydrodiols, that c) – R,R-dihydrodiols should be preferentially formed from R,S-arene oxides as well as d) – the preferential formation of *trans*-dihydrodiols from R,R-dihydrodiols with 'bay-region' double bonds. Until now the model was demonstrated to be also valid in predicting the stereoselective metabolism of polycyclic aromatic hydrocarbons like phenanthrene, chrysene and benzanthracene[19,23].

Numerous drugs are known to be metabolized, at least partly, by this cytochrome P-450, e.g. phenacetine, propranolol and warfarin[24]. A series of elaborate reviews on stereoselective metabolism of drugs by the mixed-function oxidase system have appeared in recent years[1–4]. Models as developed for polycyclic aromatic hydrocarbons are not as yet available for drugs and therefore it is difficult to predict stereoselectivity in their oxidative biotransformation.

Stereoselectivity in epoxide hydrolases

Epoxide hydrolases are enzymes which catalyze the hydrolysis of alkene oxides into diols and of arene oxides into dihydrodiols (Fig. 7; [25]). Although specific examples are known in which bioactivation was evident[13], epoxide hydrolases generally inactivate electrophilic epoxides formed during metabolism[25]. Epoxides have been identified as metabolites of a wide variety of substrates possessing olefinic or

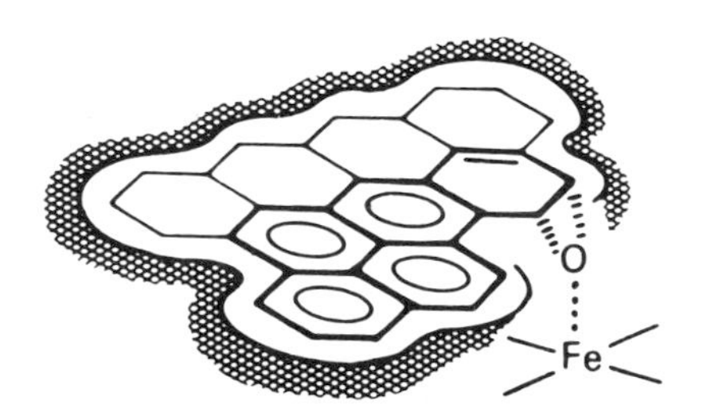

Figure 6. *Nine-membered ring model for the substrate binding site of cytochrome P450c as proposed by Jerina* et al.[19]. *Benzo(a)pyrene is drawn bound to it in such a way that benzo(a)pyrene-9S,10R-oxide is obtained. By turning benzo(a)pyrene in the model it is possible to rationalize also the stereoselective formation of benzo(a)-pyrene-(4S,5R)-oxide and benzo(a)pyrene-(7R,8S)-oxide (taken from Jerina* et al.[19]*).*

aromatic groups, such as endogenous steroids[26], environmental xenobiotics[27] as well as clinically used drugs, such as tricylic antidepressants[28], allyl- and vinyl-substituted barbiturates[29], phenylsuccinimides, carbamazepine, phenylhydantoins etc.[30].

Meanwhile, two different epoxide hydrolases have been traced, namely a microsomal membrane bound[25] as well as a cytosolic soluble one[31]. Substrate selectivities of both hydrolases is large and largely overlapping, although the cytosolic hydrolase seems to possess a larger activity towards aliphatic epoxides[31]. Mechanistically, the enzymatic hydration of epoxides most probably proceeds according to a base-catalyzed nucleophilic addition of 'activated water'[32]. A simple electrophilic (acid-

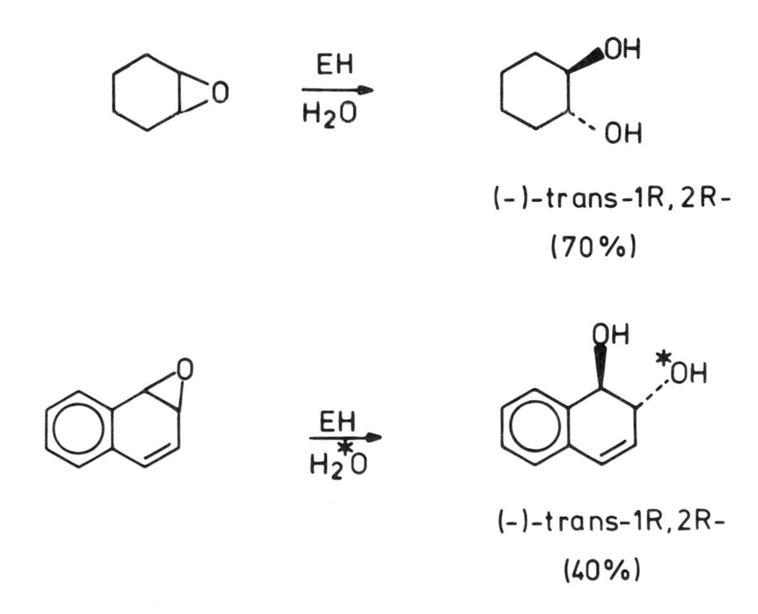

Figure 7. *Stereo- and enantioselective formation of* trans-*diols by epoxide hydrolase (EH) from cyclohexeneoxide and naftalene-oxide. Enzymatic hydrolysis of naftalene-oxide in a* $^{18}OH_2$*-medium (*$H_2{*}O$*) resulted regiospecifically in the introduction of* ^{18}O *at the 2-position (taken from Jerina* et al.[33]*).*

catalyzed) mechanism was ruled out by experiments with $^{18}OH_2$, which demonstrated regiospecific hydration at the 2-position of unlabelled styrene- and nafthaleneoxides[32,33,34].

Stereochemical studies have shown that epoxide hydrolases hydrate epoxides of cyclic olefines including arene oxides, in such a way that *trans*-diols with a varying degree of enantiomeric homogeneity are formed (Fig. 7; [33,35]). Geometrical isomers of aliphatic epoxides have also been shown to undergo *trans*-hydration, which means that *trans*-epoxides yield *erytho*-glycols and *cis*-epoxides *threo*-glycols (Fig. 8)[31].

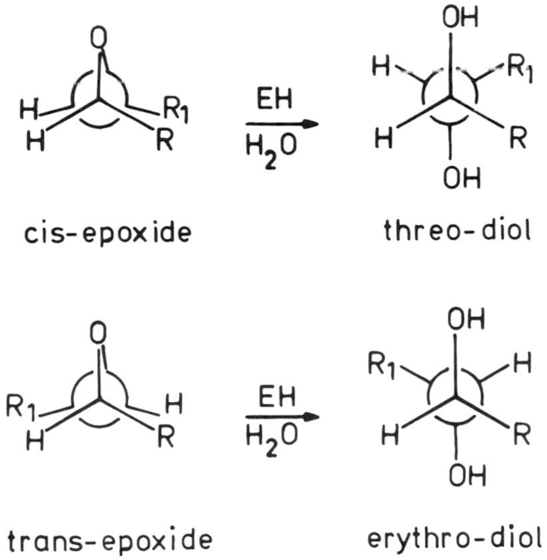

Figure 8. *Stereoselective hydrolysis of geometrical (*cis- *and* trans-*) aliphatic epoxides by epoxide hydrolase (EH) (derived from Hammock* et al.[31]*).*

In the literature low and high substrate enantioselectivity have been described for microsomal epoxide hydrolases. For example, for 4,5-benzo(a)pyrene oxide a biphasic first-order kinetic behaviour reflecting a 40-fold difference in the rates of hydration of the (+)- and (−)enantiomers was observed under certain *in vitro* conditions[36]. However, it is not possible to make generalizations with regard to substrate- or product-stereoselectivity towards structurally different epoxides[32,35]. An understanding of the mechanism of hydration of epoxides is crucial to rationalize substrate-selectivity and enzyme inhibition, both phenomena being important in the dual role of this enzyme in the metabolism of drugs and other xenobiotics.

Stereoselectivity in glutathione-transferases

In drugs and other xenobiotics possessing an electrophilic center (e.g. an easily leaving group), biotransformation reactions often include conjugation with glutathione. The enzymatic conjugation of the optically active tripeptide glutathione (GSH; glu-cys-gly, Fig. 9)[37] to a wide variety of xenobiotic substrates[38,39] as well as to some endogenous compounds is catalyzed by a family of closely related cytosolic or microsomal GSH-transferases with different but overlapping second-substrate selectivity[38,40]. Enzymatic (and also spontaneous) conjugation of drugs or other xenobiotics generally protects the organism against harmful or as a result of biotransformation potentially harmful alkylating compounds. In a limited number of

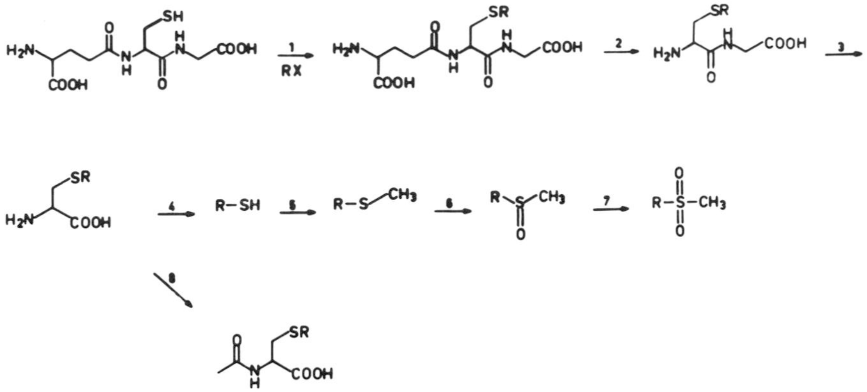

Figure 9. *Conjugation of electrophilic substrates (RX) to glutathione and subsequent metabolic reactions occurring* in vivo. *Mercapturic acids (route 8) are the major endproducts to be excreted in urine (taken from Vermeulen* et al.[37]*).*

cases, e.g. vicinal dihalogen compounds like 1,2-dibromo-[41] and 1,2-dichloroethanes[42], the conjugation to GSH gives rise to reactive 2-halogenothioethers, which are responsible for mutagenic effects. GSH transferases, like other biotransformation enzymes, thus play a dual role in the biotransformation of xenobiotics[15].

Although there is evidence for cysteine-conjugates and other sulfurcontaining metabolites such as methylthioethers, sulfones and sulfoxides (Fig. 9)[37] it is generally believed that *in vivo*, mercapturic acids are the major end-products of the glutathione pathway. During the last few years the measurement of mercapturic acids in urine has become a major tool in studying glutathione conjugation *in vivo*. On the one hand this is because the excretion of these products in urine may reflect the exposure of a living organism to (potentially) electrophilic compounds[43], and on the other hand because their structure reflects the stereo- and regioisomeric mechanisms of GSH conjugation *in vivo*[15,37].

With regard to the stereoselectivity of GSH conjugation to rigid epoxides it was, for example, found by identifying and quantifying mercapturic acids excreted in urine of rats treated with the model compound cyclohexene oxide, that only two *trans*-2-hydroxycyclohexyl-mercapturic acids and no *cis*-isomers were *in vivo* found (Fig. 10)[44,45]. This observation is an accordance with an enzymatically catalyzed

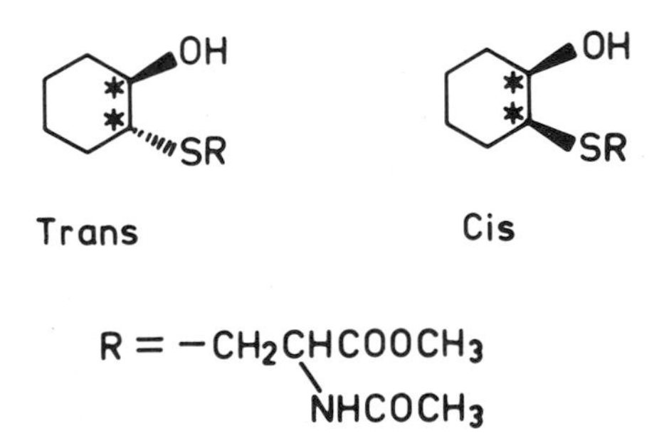

Figure 10. *Chemical structures of* trans- *and* cis-*2-hydroxycyclohexyl-mercapturic acids. Only two* trans-*diastereoisomers are formed from cyclohexeneoxide by the rat (taken from Vermeulen* et al.[45]*).*

direct attack of the nucleophilic sulfur atom of GSH on the electrophilic epoxide according to a Sn_2-substitution mechanism. A stereoselective excretion of the two *trans*-diastereoisomeric mercapturic acids in dependence of the dose of cyclohexene oxide was suggested to be due to the involvement of different isoenzymes of GSH-transferases[44].

In vitro, a similar *trans*-addition mechanism was found for the enzymatic conjugation of (±)-benz(a)pyrene-4,5-oxide using rat liver cytosol as a source of transferase activity[46]. Apparently, both positional isomers (at C_4 and C_5; Fig. 11)[47] were produced, however, with a preference for two diastereoisomers. Subsequent investigations with optically active (+)-4S,5R and (-)-4R,5S-benzo(a)pyrene-4,5-oxides[47] have shown that the rat liver cytosol GSH-transferases preferentially catalyze the formation of the 5-gluthionyl-isomer (4S,5S) from the (+) and the 4-gluthionyl-isomer (4S,5S) from the (-)-benzo(a)pyrene-4,5-oxide isomer. Apparently, these GSH-transferases prefer to attack the oxirane carbon of the benzo(a)pyrene-4,5-oxide, which has the absolute R-configuration[47].

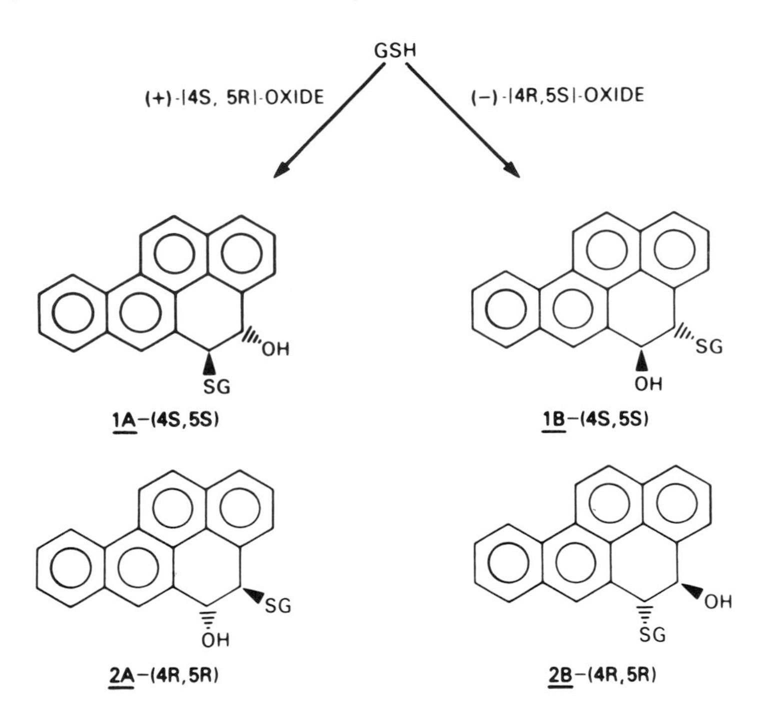

Figure 11. *Structures of four possible glutathione (GSH) conjugates from (+)- and (-)-benzo(a)pyrene-4,5-oxide. Cytosolic rat liver GSH-transferases preferentially form the conjugates IA and IB from racemic benzo(a)pyrene-4,5-oxide (taken from Armstrong* et al.[47].

Apart from aliphatic and aromatic epoxides, GSH-transferases catalyze the conjugation of GSH to many other hydrophobic substrates possessing leaving groups, such as phosphates, sulphates and halogenides[38,39]. Regarding the enzymatic reaction in the latter class of substrates, rat liver cytosolic GSH transferases were found to conjugate chiral α-phenylethyl-chloride and -bromide with a high degree of substrate-stereoselectivity[48]. S-enantiomers were far better substrates than R-enantiomers (Fig. 12)[48]. The conjugation proceeded with complete conversion of configuration at the benzylic carbon atom, which is consistent with a Sn_2-type of

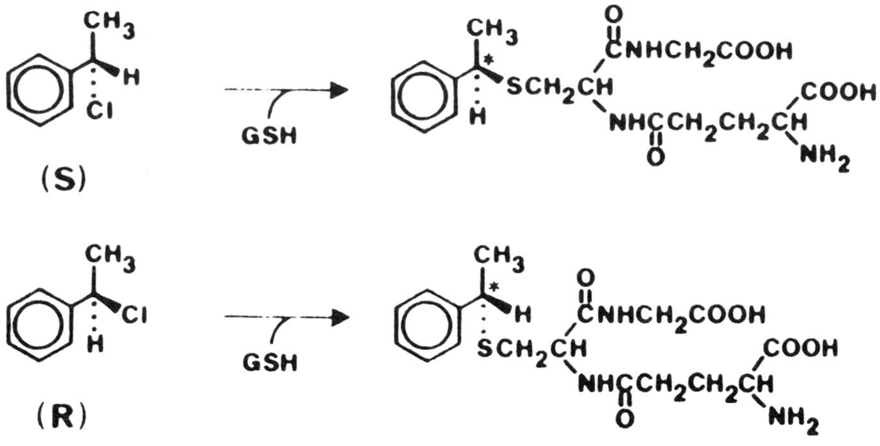

Figure 12. *Stereochemical course of the conjugation of glutathione to (S)- and (R-)-phenylethylchlorides. Cytosolic rat liver GSH-transferases preferentially react with the (S)-enantiomer, according to a S_{n2}-type mechanism (derived from Mangold* et al.[48]*).*

substitution mechanism involving a direct attack of the sulfhydryl group of GSH at the electrophilic benzylic carbon atom[48].

Although a number of drugs and related substances, e.g. phenacetine, nitroglycerine, ethacrynic acid, estradiol and bromosulfophtaleine, are known to be partly conjugated to GSH, far larger numbers of other electrophilic xenobiotics (environmental chemicals and pesticides) have been shown to be substrates for this type of conjugation[39,49,50].

Stereoselectivity in glucuronyltransferases

β-D-Glucuronic acid is an important optically active cofactor for the microsomal UDP-glucuronyltransferases, which catalyze the conjugation of this glucuronic acid to O-, N- or C-functional groups containing substrates[51].

Substrate-stereoselectivity for this enzyme-system has been shown to occur in the metabolism of hexobarbital[52] and propranolol[53,54]. In the case of propranolol species differences in the stereoselectivity of glucuronidation of propranolol itself became apparent (Fig. 13, Table 1). 4′-Hydroxypropranolol did not seem to be conjugated in a stereoselective manner[53]. Recently, however, Walle *et al.*[54] reported the identification of three new ring-hydroxylated metabolites of propranolol, *viz* 2′-5′- and 7′-hydroxypropranolol. The (+)-enantiomer of latter metabolite was found to be formed stereospecifically from (+)-propranolol and in contrast to (–)-7-hydroxypropranolol suggested to be preferentially glucuronidated before excretion into urine.

The underlying mechanisms of stereoselective glucuronidation reactions are not yet clearly understood. It is important, however, to realize possible stereoselective effects in the formation of precursor-metabolites as well as in their further metabolism or excretion.

Metabolic inversion of configuration

An interesting but rather rarely observed phenomenon in the metabolism of xenobiotics is inversion of configuration as the result of a metabolic reaction. A well documented example of such a phenomenon is the interconversion of (epimeric or)

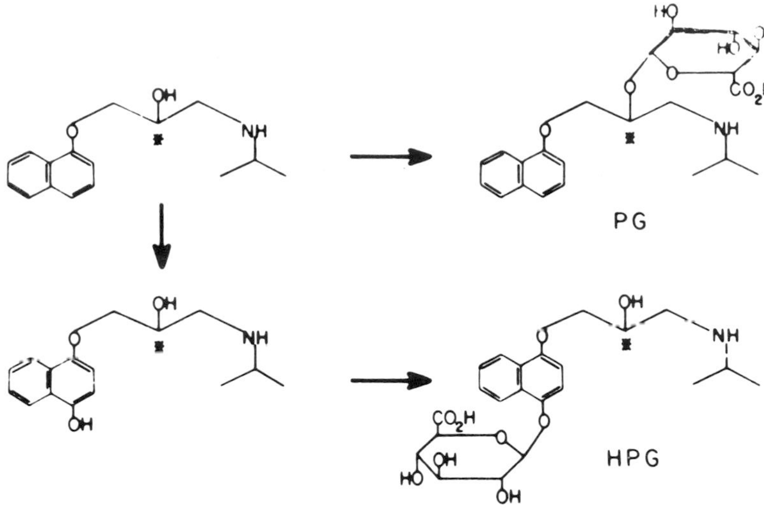

Figure 13. *Metabolic scheme for the formation of 4'-hydroxypropranolol and the glucoronide conjugates of propranolol (PG) and 4'-hydroxypropranolol (HPG) (taken from Thompson* et al.[53]*).*

Table 1. *Stereoselective glucuronidation*

	Propranolol-3-O-glucuronide	
Species	Ratio R-(+)/S-(-)	Reference
Human (urine)	1 : 1	Silber & Riegelman[55]
Dog (urine)	1 : 2	Ehrsson[56]
Rabbit (purified liver GT)	2 : 1	Fenselau & Johnson[57]
Rat (liver microsomes)	2 : 1	Thompson et al.[53]

enantiomeric cycloalkanols[58]. Mechanistically this metabolic reaction was explained by a reversible enzymatic alcohol-alkanone oxido-reduction reaction. The position of the equilibrium of this reversible reaction is probably dependent on the ratio of the necessary cofactors NAD/NADH for the alcohol-dehydrogenase involved[59].

Ibuprofen (Fig. 14) is an anti-inflammatory drug for which metabolic substrate-as well as product-stereoselective inversion was observed in several species[60], including man[61]. R-(-)-Ibuprofen, which is the pharmacologically far lesser active isomer, was found to be excreted in urine as its active S-(+)-enantiomer or as mainly dextrorotatory hydroxy- and carboxy-metabolites (Fig. 14, Table 2). Similar observations were made in studies on the disposition of R-(-)-α-methylfluorene-2-acetic acid (cycloprofen)[62] and R-(-)-4-chlorophenyl)-5-benzoxazol-2-propionic acid (benoxaprofen)[63]. Except metabolic inversion, the latter drug is not known to be further metabolized by animals or man. Results obtained after oral and intravenous administration of R-(-)-benoxaprofen to rats, as well as *in vitro* experiments with rat gut microflora in inverted intestinal sac preparations, suggest that this kind of configurational inversion reactions *in vivo* are not taking place in the liver, but instead by a R-(-)-arylproprionic acid isomerase-enzyme in the gut wall[60,63].

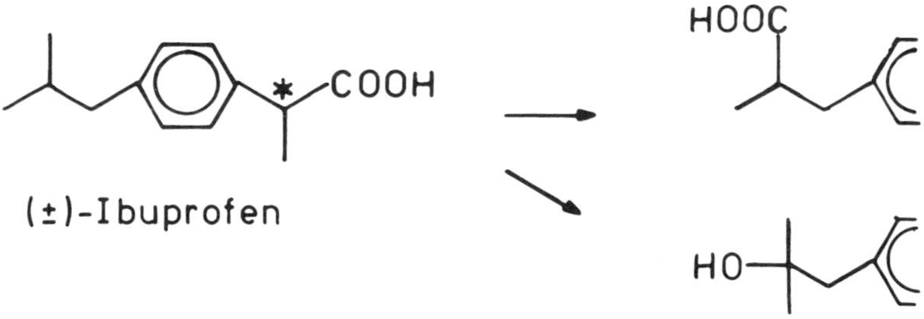

HOOC

(±)-Ibuprofen

HO

Figure 14. *Chemical structure and part of the metabolic fate of ibuprofen (derived from Kaiser* et al.[61]*).*

Table 2. *Inversion of ibuprofen in humans*[a]

	Administered S/R-ratio	Excreted S/R-ratio
S-(+)	95 : 5	95 : 5
R-(-)	6 : 94	80 : 20
Racemic	50 : 50	70 : 30

[a] Adapted from Kaiser et al.[61]

Stereoselectivity in hexobarbital disposition

Hexobarbital is a chiral compound which was introduced as a hypnotic drug in 1932. It is a highly lipophilic drug and it has also widely been used as an i.v. anaesthetic agent. At present hexobarbital is mainly used as a model substrate in drug metabolism studies. Sleeping times, elimination half-lives or clearances of the drug are often employed as parameters reflecting oxidative hepatic enzyme activity[29]. As is indicated in Fig. 15, there exist two major oxidative metabolic pathways for hexobarbital, *viz* allylic oxidation, leading to the formation of 3'-hydroxy- and 3'-ketohexobarbital[64] and the recently identified epoxidation-pathway, ultimately leading to the formation of 1,5-dimethylbarbituric acid[29].

In 1965, Knabe *et al.* described the separation of racemic hexobarbital in its enantiomers[65]. The enantiomers were subsequently shown to exhibit considerable differences in pharmacological effects. A considerably larger hypnotic and anaesthetic potency is to be attributed to S-(+)-hexobarbital in several species[66].

In vitro studies with hepatic microsomes elucidated on the one hand a pronounced difference in the rate of oxidative metabolism of the enantiomers; in rats and mice the S-(+)-enantiomer was metabolized considerably faster than the R-(-)-enantiomer, which could mainly be ascribed to a lower K_m-value[67-69]. On the other hand there was also a significant gender and species difference in the metabolism of the hexobarbital-enantiomers which was already referred to in this chapter[67-69]. These observations as well as differential effects of enzyme inducers like phenobarbital and 3-methylcholanthrene and inhibitors like SKF-525A, have led to the suggestion that different forms of cytochrome P-450 are involved in the metabolism of the enantiomers of hexobarbital[68,69].

Stereoselective degradation of the enantiomers of hexobarbital leads to a predominant formation of α-hydroxy-(+)-hexobarbital from S-(+)-and β-hydroxy-(-)-hexobarbital from R-(-)-hexobarbital, as was demonstrated in rat liver microsomes[68]. Besides, 3'-hydroxy-derivatives bearing a levorotatory 3'-S-configu-

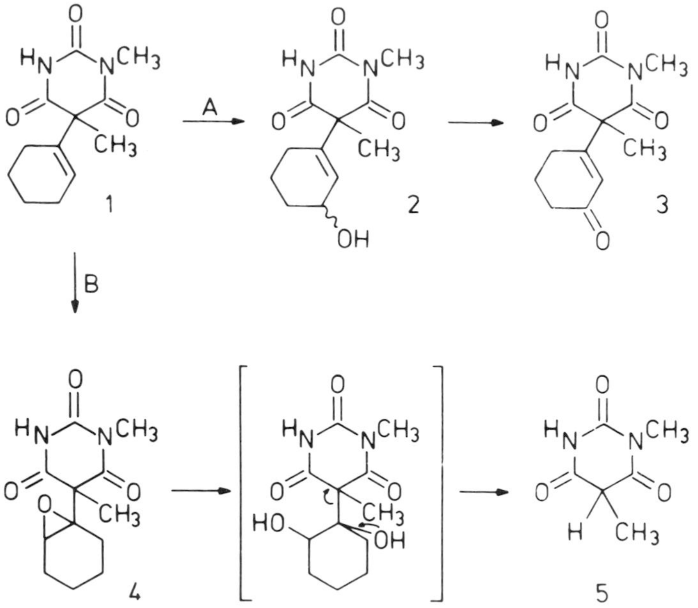

Figure 15. *Metabolic fate of hexobarbital*[1]. *The major pathways are the allylic oxidation (A) leading to 3'-hydroxy-*[2] *and 3'-ketohexobarbital*[3] *and the epoxide-diol pathway*[29], *which leads to 1',2'-epoxyhexobarbital*[4] *and 1,5-dimethylbarbituric acid*[5] *(taken from Vermeulen*[29]*).*

ration were shown to be preferentially dehydrogenated to the respective 3'-ketometabolites by a cytosolic dehydrogenase of rabbits and guinea pigs[69]. In rabbits it was also established that the conjugation of α-hydroxy-hexobarbital from the R-(–)- and β-hydroxy-hexobarbital from S-(+)-hexobarbital to glucuronic acids, i.e. glucuronidation of hydroxy-hexobarbitals with a levorotatory S-configuration around the 3'-carbon atom, occurs preferentially[52].

In recent years several *in vivo* studies have been conducted on the pharmacokinetics of the enantiomers. Except humans[8] all other species investigated, e.g. mice, rabbit, guinea pig, metabolized S-(+)-hexobarbital more rapidly than the R-(–)-enantiomer. This was also the case in rats[70,71]. The difference in intrinsic clearances between S-(+)- and R-(–)-hexobarbital, as reflected by the difference in the areas under their respective blood concentration time curves after oral administration to the rat (Fig. 16), was about sevenfold[71]. This difference was in the order of magnitude as to be expected on the basis of V_{max} - and K_m- values estimated in rat liver microsomes. It was, however, far larger than the difference between the corresponding elimination half-lives or systemic clearances as estimated upon intravenous administration of the isomers. The observed discrepancies are to be attributed to hepatic bloodflow limited metabolism of hexobarbital[29,72], a phenomenon which was further substantiated by the relatively high extraction ratios found on oral administration of the enantiomers to the rat, *viz* extraction ratio E = 0.64 for R-(–)- and 0.95 for S-(+)-hexobarbital[71]. As a consequence of the high and stereoselective first-pass metabolism of the enantiomers of hexobarbital changes in hepatic enzyme activity are not well reflected anymore by changes in elimination half-life, systemic clearance nor sleeping time of racemic hexobarbital.

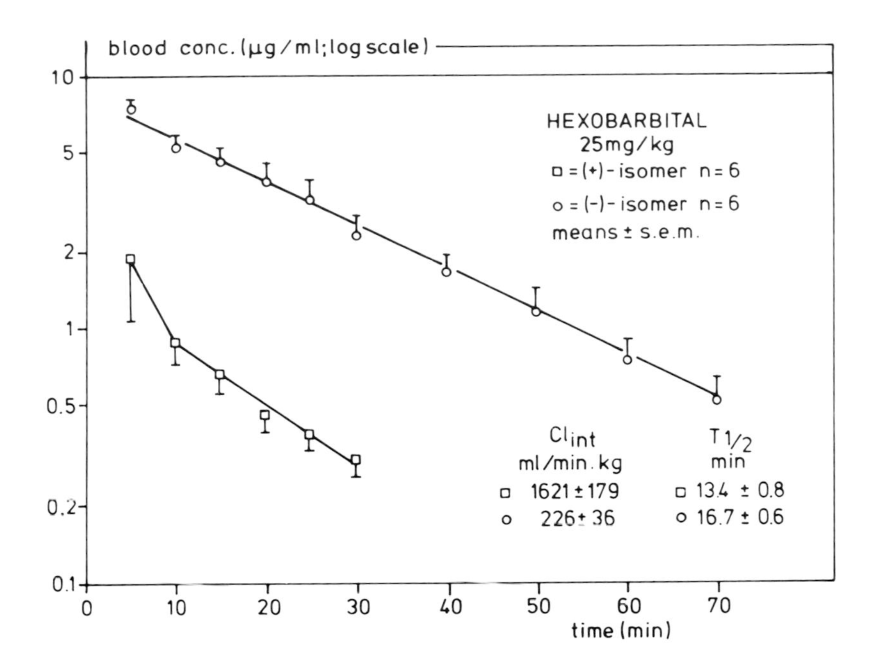

Figure 16. *Mean blood-concentration time profiles of (+)- and (–)-hexobarbital in six different rats obtained after oral administration of 25 mg/kg of the enantiomers (taken from Van der Graaff* et al.[71]*)*.

Our present research is dealing with the question whether in the view of the use of hexobarbital as a model substrate to assess oxidative hepatic drug metabolizing capacity, it would be preferable to use the pharmacologically active S-(+)-enantiomer when sleeping times are used or to employ the inactive R-(–)-isomer when intrinsic clearances are used as parameters to reflect hepatic enzyme activity.

Analytical techniques to study stereoselective drug disposition

A frequently encountered problem in drug metabolism and pharmacokinetic studies is to find an analytical technique that enables to discriminate between stereoisomeric drugs or metabolites present in low concentrations in biological fluids. In principle stereoisomers can show either an enantiomeric or diastereomeric relationship. Diastereoisomers, being stereoisomers which are not mirror images, differ in their physicochemical properties and can thus be separated or distinguished by chromatographic, spectroscopic or other techniques. Enantiomers, however, being not superimposable mirror images (differing only in optical rotation) can be distinguished solely by techniques measuring directly optical activities or by interaction of the enantiomers with chiral compounds, thus becoming diastereomeric compounds.

A valuable review on stereochemical methodology in drug metabolism was published some years ago by Testa and Jenner[73].

Separation of diastereoisomers

The separation of diastereoisomeric compounds (e.g. epimers) can be based on a proper utilization of structure-related differences like electronic distribution, solubility, volatility or reactivity etc. This approach has been followed in many recent studies[73-75]. Usually chromatographic techniques like thin layer-, gas- and liquid chromatography or spectroscopic techniques like ^{1}H- or ^{13}C-magnetic resonance are employed. However, systematic studies on the analysis of diastereoisomers, in particularly in the area of drug metabolism are still rare.

Discrimination between enantiomers

The separation between enantiomers is generally more complicated, since enantiomers solely differ in optical activity. Polarimetric measurement of optical rotation (milligram amounts of pure substances), usually carried out at a wave length of the sodium D-line (589 nm), has permitted the study of stereoselective hydroxylation reactions in e.g. ethylbenzene and indane *in vitro*[73]. Optical rotation dispersion (ORD) and circular dichroism (CD) techniques (spectropolarimeters), allow the determination of enantiomeric percentages as well as the elucidation of absolute configurations in e.g. (w-1)-hydroxylation reactions of barbiturates or binding of substrates to metabolizing enzymes like cytochrome P-450[73].

A different approach to the separation of enantiomers is based on the use of chromatographic techniques such as gas- and liquid chromatography. In principle these techniques are advantageous for metabolism studies since they allow the separation and quantitative determination of sub-microgram amounts of compounds present in biological fluids. There exist two principally different possibilities in resolving optical isomers with GLC and HPLC. The first one is derivatization of enantiomers with suitable chiral reagents, subsequently followed by separation of the resulting diastereoisomers with normal stationary phases[74,75]. This technique was recently employed in a stereospecific assay of blood concentrations of (+)-and (–)-propranolol in humans and dogs[55]. Derivatization of the secondary amine-group of propranolol-enantiomers for this purpose was carried out with N-trifluoroacetyl-(–)-prolylchloride (TPC; Fig. 17). The second possibility is a direct separation of enantiomers by the use of chiral stationary phases forming *in situ* diastereomeric association complexes with enantiomers. Practical applications of a commercially available chiral stationary phase for HPLC, which was prepared by ionically binding R-N-3,5-dinitrobenzoylphenylglycine to an achiral support, were recently reviewed[76]. Similarly, chiral stationary phases have been become available for gas chromatography, e.g. Chirasil-Val (L-valine-tert-butyl-amide-carboxyalkyl-methyl-siloxane;[77] and XE-60-S-valine-S-phenylethylamide)[78]. Most gas chromatographic applications described until now are dealing with the separation of amino acids, amino alcohols, amines and hydroxy acids[78], although a limited number of drugs like ephedrine and dopa were analysed[77]. In Fig. 18 a gas chromatogram is presented showing the separation of racemic hexobarbital into its enantiomers on a SCOT-capillary column coated with Chirasil-Val. This system was recently used in a study on the kinetics of the isomers of hexobarbital upon oral and intravenous administration of racemic hexobarbital to rats[79].

At present the major problems complicating the use of chiral stationary phases are their limited long-term and, particularly in the case of GLC, limited temperature stability (*viz* 200°C).

With regard to the use of magnetic resonance spectroscopy, a diastereomeric relationship is also required to discriminate between enantiomers. This can be achieved either by chemical derivatization with optically active reagents or simply by intermolecular interaction between chiral shift reagents, such as lanthanide-derivatives and the enantiomers[73].

A relatively new but rather promising technique for routine analysis of enantiomeric drugs or metabolites in biological fluids, involves immumo-assays with enantioselective antisera. Such antisera were shown to be useful in studies on

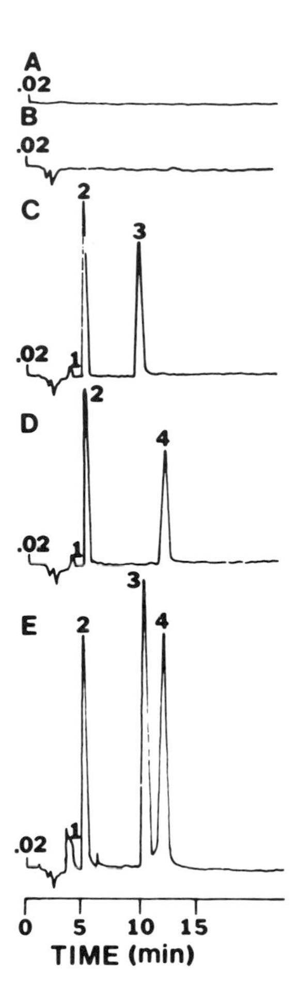

Figure 17. *Stereospecific assay of propranolol: HPLC chromatograms resulting from derivatization with N-trifluoroacetyl-(–)-prolylchloride (TPC) of control dog (trace A) and human (trace B) plasma, and human plasma spiked with 100 ng/ml of (–)-propranolol (trace C, peak 3), (+)-propranolol (trace D, peak 4) and (±)-propranolol (trace E, peak E). The internal standard is N-ethyl-propranolol (peak 2) (taken from Silber* et al.[55]*).*

the stereoselective disposition of enantiomeric drugs like propranolol[80], warfarin[81] as well as the antimalaria agent WR-171.699[82]. The generation of stereospecific antisera, however, is certainly not yet routine practice.

Hitherto it is common practice that asymmetric drugs are administered as racemates, even though the enantiomers have different pharmacological activities, metabolism or pharmacokinetics. It is therefore very important to dispose of techniques enabling the specific determination of stereoisomeric drugs or metabolites.

Conclusions

Stereoselective metabolism of drugs and other xenobiotics is very important, not only from a quantitative but also from a pharmacological and toxicological point of view. Elucidation of possible substrate- and product stereoselective effects in the

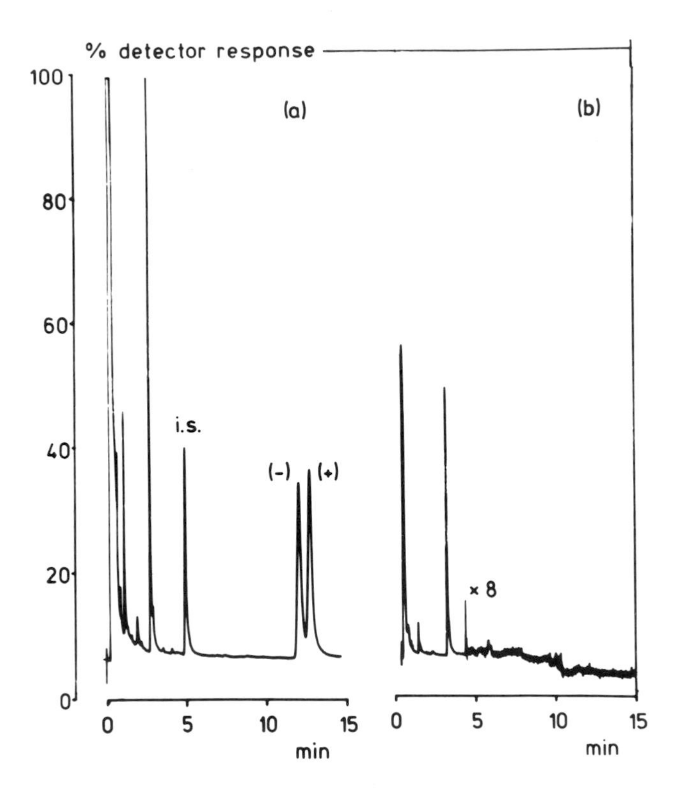

Figure 18. *Gas chromatograms of rat blood, spiked with 10.2 ng/ml (±)-hexobarbital (panel a) and blank rat blood (panel b), obtained with a capillary chirasil-Val column and a nitrogen selective detector. The internal standard (i.s.) is phensuccimide (taken from Van der Graaff* et al.[79]*).*

disposition of drugs should be pursued at an early stage of drug development, in order to enhance their efficacy in those cases where it concerns stereoisomers with significantly different pharmacological or (potentially) toxicological effects.

The stereoselective action of drug and xenobiotic metabolizing enzymes, except epoxide hydrolases and glutathione transferases, is difficult to generalize. With regard to enantio- (product-) selectivity the lack of predictability holds true for all metabolizing enzymes. Probably the existence of multiple enzyme forms with overlapping substrate-selectivities in one enzyme system is the most likely explanation for this. As a consequence, it is difficult to predict the stereoselective disposition of drugs and other xenobiotics in general, or even to predict stereoselective effects observed in one drug to another closely related structural analogue.

However, it also should be realized that fundamental studies with stereoisomers of drugs and other xenobiotics may furnish information about the space and geometry at and near the active sites of metabolizing enzymes and as such about underlying mechanisms of stereoselective metabolism. Recent developments in analytical techniques enabling the simultaneous determination of enantiomers, e.g. GLC and HPLC, will be helpful in pursuing this objective.

References

1. P. Jenner and B. Testa, *Drug Metab. Rev. 2,* 117 (1973).
2. P.N. Patil, D.D. Miller and U. Trendelenburg, *Pharmacol. Rev. 26*, 223 (1975).

3. B. Testa and P. Jenner, *In* 'Concepts in Drug Metabolism', Part A, p. 75-143 (P. Jenner and B. Testa, eds.). Marcel Dekker Inc., New York, 1980.
4. B. Testa, *Pharm. Int. 2,* 34 (1981).
5. G.P. Jackman, A.J. Mclean, G.L. Jennings and A. Bobik, *Clin. Pharmacol. Ther. 30,* 291 (1981).
6. P.J. Murphy, R.C. Nickander, G.M. Bellamy and W.L. Kurtz, *J. Pharmacol. Exp. Ther. 199,* 415 (1976).
7. A.S. Nies, G.H. Evans and D.G. Shand, *J. Pharmacol. Exp. Ther. 184,* 716 (1973).
8. D.D. Breimer and J.M. van Rossum, *J. Pharm. Pharmacol. 25,* 762 (1973).
9. J.L. Holzman and J.A. Thompson, *Drug Metab. Disp. 3,* 113 (1975).
10. R. Kato, *Pharmacol. Ther. 6,* 41 (1979).
11. J.M. van den Broek, *Ph.D.-thesis* 'Species Differences in the Induction of Drug Metabolism by Rifampicin'. University of Leiden, The Netherlands, 1981.
12. P.J. van Bladeren, A. van der Gen, D.D. Breimer and G.R. Mohn, *Biochem. Pharmacol. 28,* 2521 (1979).
13. P. Bentley, F. Oesch and H. Glatt, *Arch. Toxicol. 39,* 65 (1977).
14. A.Y.H. Lu and S.B. West, *Pharmacol. Rev. 31,* 177 (1980).
15. P.J. van Bladeren, *Ph.D.-thesis* 'The Dual Role of Glutathione Conjugation in the Biotransformation of Xenobiotics'. University of Leiden, The Netherlands, 1981.
16. V. Ullrich and P. Kremers, *Arch. Toxicol. 39,* 41 (1977).
17. V. Ullrich, *In* 'Microsomes and Drug Oxidations' (V. Ullrich *et al.,* eds.). Perganon Press, New York, 1977.
18. D.R. Thakker, H. Yagi, H. Akagi, M. Koreeda, A.Y.H. Lu, W. Levin, A.H. Wood, A.H. Conney and D.M. Jerina, *Chem. Biol. Interactions* 16, *281* (1977).
19. D.M. Jerina, D.P. Michand, R.J. Feldman, R.N. Armstrong, K.P. Vyas, D.R. Thakker, H. Yagi, P.E. Thomas, D.E. Ryan and W. Levin, *In* 'Fifth International Symposium on Microsomes and Drug Oxidations', p. 195 (R. Sato and K. Kato, eds.). Japan Scientific Societies Press, 1982.
20. R.N. Armstrong, W. Levin, D.E. Ryan, P.E. Thomas, H. Duck Mah and D.M. Jerina, *Biochem. Biophys. Res. Commun. 100,* 1077 (1981).
21. K.M. Straub, T. Meehan, A.L. Burlingame and M. Calvin, *Proc. Natl. Acad. Sci. U.S.A. (Biochem). 74,* 5285 (1977).
22. M.R. Osborne, *Trends in Biochem. Sci. 9* (1979).
23. P.J. van Bladeren, R.N. Armstrong, D. Cobb, D.R. Thakker, D.E. Ryan, P.E. Thomas, N.D. Sharma, D.R. Boyd, W. Levin and D.M. Jerina, *Biochem. Biophys. Res. Commun. 106*, 602 (1982).
24. G.W. Dawson and R.E. Vestal, *Pharmac. Ther. 15,* 207 (1982).
25. F. Oesch, *In* 'Progress in Drug Metabolism', Part 3, p. 253 (J.W. Bridges and L.F. Chasseaud, eds.). Wiley, Chichester, 1979.
26. B. Kadis, *J. Steroid. Biochem.* 9, *75* (1978).
27. A.Y.H. Lu, D.M. Jerina and W. Levin, *J. Biol. Chem. 252,* 3715 (1977).
28. A. Frigerio, M. Cavo-Briones and G. Belvedere, *Drug Metab. Rev. 5,* 197 (1976).
29. N.P.E. Vermeulen, *Ph.D-thesis* 'The Epoxide-diol Pathway in the Metabolism of Hexobarbital and Related Barbiturates'. University of Leiden, The Netherlands, 1980.
30. F. Oesch, *Biochem. Pharmacol. 25,* 1935 (1976).
31. B. Hammock, M. Ratcliff and D.A. Schooley, *Life Sci. 27,* 1635 (1980).
32. R.P. Hanzlik, M. Edelman, W.J. Michaely and G. Scott, *J. Am. Chem. Soc. 98,* 1952 (1976).
33. D.M. Jerina, H. Ziffer and J.W. Daly, *J. Am. Chem. Soc. 92,* 1056 (1970).
34. G.C. Dubois, E. Appella, W. Levin, A.Y.H. Lu and D.M. Jerina, *J. Biol. Chem. 253,* 2932 (1978).
35. R.N. Amstrong, B. Kedzierski, W. Levin and D.M. Jerina, *J. Biol. Chem. 256,* 4726 (1981).
36. R.N. Amstrong, W. Levin and D.M. Jerina, *In* 'Microsomes, Drug Oxidations and Chemical Carcinogenesis' (M.J. Coon, A.H. Conney, R.W. Estabrook, N.V. Gelboin, J.R. Gillette and P.J. O'Brien, eds.). Academic Press Inc. New York, 1980.
37. N.P.E. Vermeulen, W. Onkenhout and P.J. van Bladeren, *In* 'Chromatography and Mass Spectromy in Biomedical Sciences' (A. Frigerio ed.). Elsevier Scientific Publishing Co., Amsterdam, 1982.
38. L.F. Chasseaud, *Drug. Metab. Rev. 2,* 185 (1973).

39. I.M. Arias and W.B. Jacoby, eds. 'Glutathione, Metabolism and Function. Raven Press, New York, 1976.
40. A.J. Baars and D.D. Breimer, *Ann. Biol. Clin. 38*, 49 (1980).
41. P.J. van Bladeren, D.D. Breimer, G.M.T. Rotteveel-Smijs, P. de Knijf, G.R. Mohn, P. van Meeteren-Wälchli, W. Buijs and A. van der Gen, *Carcinogenesis 2,* 499 (1981).
42. U. Rannug, A. Sundvall and C. Ramel, *Chem. Biol. Int. 20,* 1 (1978).
43. R. van Doorn, Ch.M. Leyendekkers, R.P. Bos, R.M.E. Brauns and P.Th. Henderson, *Ann. Occup. Hyg.* 24, 77 (1981).
44. P.J. van Bladeren, D.D. Breimer, J.A.T.C.M. van Huijgenvoort, N.P.E. Vermeulen and A. van der Gen, *Biochem. Pharmacol. 30,* 2499 (1981).
45. N.P.E. Vermeulen, J. Cauvet, W.C.M.M. Luijten and P.J. van Bladeren, *Biomed. Mass Spectrom. 7,* 413, 1980.
46. O. Hermandez, M. Walker, R.H. Cox, G.L. Foureman, B.R. Smith and J.R. Bend, *Biochem. Biophys. Res. Commun. 96,* 1494, 1980.
47. R.N. Armstrong, W. Levin, D.E. Ryan, P.E. Thomas, H. Duck Mah and D.M. Jerina, *Biochem. Biophys. Res. Commun. 100,* 1077 (1081).
48. J.B. Mangold and M.M. Abel-Monem, *Biochem. Biophys. Res. Commun. 96,* 333 (1980).'
49. P.L. Grover, *In* 'Drug Metabolism - from Microbe to Man', p. 105 (D.V. Park and R.L. Smith, eds.). Taylor and Francis, London, 1977.
50. B. Mannervik, C. Guthenberg, I. Jakobson and M. Warholm, *In* 'Conjugation Reactions in Drug Biotransformation', p. 101 (A. Aitio, ed.). Elsevier, Amsterdam, 1978.
51. A. Aitio, ed. 'Conjugation Reactions in Drug Biotransformation', chapter Nature UDP-glucuronyl, pp. 167-269. Elsevier/North-Holland Biomedical Press, Amsterdam, 1978.
52. K. Miyano, T. Ota and S. Toki, *Drug Metab. Disp. 9,* 60 (1981).
53. J.A. Thompson, J.E. Hull and K.J. Norris, *Drug Metab. Disp. 9,* 466 (1981).
54. T. Walle, J.E. Oatis, U.K. Walle and D.R. Kanpp, *Drug Metab. Disp. 10,* 122 (1982).
55. B. Silber and S. Riegelman, *J. Pharmac. Exp. Ther. 215,* 643 (1980).
56. H. Ehrsson, *J. Pharm. Pharmacol. 27,* 971 (1975).
57. C. Fenselau and L.P. Johnson, *Fed. Proc. Fed. Am. Sci. Exp. Biol. 38,* 444 (1979).
58. R. Takenoshita and S. Toki, *Biochem. Pharmacol. 27,* 989 (1978).
59. H. Kono, M. Fujii, T. Sokabe and J. Kaneshige, *Enzyme 24,* 112 (1979).
60. W.J. Wechter, D.G. Loughhead, R.J. Reisher, G.J. van Geissen and D.G. Kaiser, *Biochem. Biophys. Res. Commun.* 61, 833 (1974).
61. D.G. Kaiser, G.J. van Geissen, R.J. Reisher and W.J. Wechter, *J. Pharm. Sci. 65,* 269 (1976).
62. S.J. Lan, K.J. Kripalani, A.V. Dean, P. Egli, L.T. Difazio and E.C. Schreiber, *Drug Metab. Disp. 4,* 330 (1976).
63. R.G. Simmonds, T.J. Woodage, S.M. Duff and J.N. Green, *Eur. J. Drug Metab. Pharmacokin. 5,* 169 (1980).
64. M.T. Bush and W.L. Weller, *Drug Metab. Rev. 1,* 249 (1972).
65. J. Knabe and R. Kräuter, *Arch. Pharm.* 298, 1 (1965).
66. G. Waldström, *Eur. J. Pharmacol.* 59, 219 (1979).
67. A.P. van den Berg, *Ph.D.-thesis* 'Cytochrome P-450 Substrate Interaction and Hepatic Drug Metabolism in the Mouse', University of Rotterdam, The Netherlands, 1977.
68. K. Miyano, Y. Fujii and S. Toki, *Drug Metab. Disp. 8,* 104 (1980).
69. K. Miyano and S. Toki, *Drug Metab. Disp. 8,* 111 (1980).
70. D.D. Breimer and J.M. van Rossum, *Eur. J. Pharmacol. 26,* 321 (1974).
71. M. van der Graaf, N.P.E. Vermeulen, R.P. Joeres and D.D. Breimer, *Submitted for publication*
72. N.P.E. Vermeulen, M. Danhof, I. Setiawan and D.D. Breimer, *Submitted for publication.*
73. B. Testa and P. Jenner, *In* 'Drug Fate and Metabolism: Methods and Techniques' (E.R. Garrett and J.L. Hirtz, eds.). Marcel Dekker Inc., New York, p. 143, 1978.
74. T. Tamegai, M. Ohmae, K. Kawabe and M. Tomoeda, *J. Liq. Chromatogr. 2,* 1229 (1979).
75. B. Halpern, *In* 'Handbook of Derivatives for Chromatography', p. 104 (K. Blau and G.S. King, eds.), Heyden, London, 1977.
76. W.H. Pirkle, J.M. Finn, B.C. Hamper, J. Schreiner and J.R. Pribish, *In* 'Asymmetric Reactions and Processes in Chemistry', A.C.S. Symposium Series No. 185, p. 245 (E.L. Eliel and S. Otsuka, eds.), 1982.

77. H. Frank, G.J. Nicholson and E. Bayer, *J. Chromatogr. 146,* 197 (1978).
78. W.A. König, S. Sievers and I. Benecke, *In* 'Capillary Chromatography; 4th International Symposium, Heidelberg' (R.E. Kaiser, ed.). Institute of Chromatography, Bad Dürkheim, 1981.
79. M. van der Graaff, N.P.E. Vermeulen, A.J.M. Hoek and D.D. Breimer, *To be published.*
80. K. Kawashima, A. Levy and S. Spector, *J. Pharmacol. Exp. Ther. 196,* 517 (1976).
81. C.E. Cook, N.H. Ballentine, T.B. Seltzman and C.R. Tallent, *J. Pharmacol. Exp. Ther. 210,* 391 (1979).
82. C.E. Cook, T.P. Seltzman, C.R. Tallent and J.D. Woolen, *J. Pharmacol. Exp. Ther. 220,* 568 (1982).

Stereoselectivity and Drug Distribution

C.A.M. Van Ginneken, J.F. Rodrigues de Miranda and A.J. Beld

Abstract

Stereoselective distribution may arise as a consequence of carrier-mediated transport or of selective binding to biomacromolecules, among which plasma proteins are especially interesting from the pharmacokinetic point of view. Stereoselective carrier transport has been established to exist in the organism with respect to several nutrients and endogenous compounds, especially amino acids and hexoses (a.o. glucose). As far as drugs are concerned little information is available, but present evidence does not support the idea of substantial stereoselectivity of drug transport. Also binding of drugs to plasma or tissue proteins appears to be not stereoselective as a rule. Apart from a few exceptions drug protein binding is not significantly different for enantiomers. On the whole the macrodistribution of drugs exhibits only minor stereoselectivity. The microdistribution, i.e. at the (sub)cellular level, of drugs is known to be quite selective sometimes with regard to enantiomers (e.g. uptake or re-uptake of transmitters and drugs in synaptic vesicles or nerve granules), without having important consequences for whole body distribution. After careful examination of the possible sites where stereoselective drug distribution may arise, it is concluded that many relevant questions remain unanswered up to now but that no indications are found that stereoselective distribution might be an important factor in drug action.

Introduction*

It is remarkable to see how many safeguards the living organism has at its disposal in order to prevent a nutrient with the wrong, unnatural stereochemical con-

* The discussion on stereoisomers in this paper is limited to pure enantiomers, i.e. structures that are each others image and that do not differ in their overall physicochemical properties. Diastereoisomers such as exist for any compound with more than one centre of asymmetry should be regarded as different compounds, having different physicochemical characteristics. Therefore differences in distribution behaviour of diasteroisomers would not be remarkable at all.

figuration from reaching the biophase where the right, natural isomer would provoke its action by interaction with a receptor or would combine with an enzyme system. And even when the wrong isomer reaches the relevant biophase it will usually be unable to replace there the natural isomer effectively. Furthermore the organism appears to be able to retain important nutrients in the body in a stereoselective way, so that the unnatural isomers are rapidly excreted (or sometimes metabolized). Typical examples of this stereoselective behaviour are the hexoses, especially glucose and the amino acids, where the natural enantiomers are the D- and L-forms respectively. Gastrointestinal absorption is selective for D-glucose and L-amino acids and the same holds true for the transport through the various barriers within the organism, especially those in liver and brain. In the kidney mechanisms are operating which selectively reabsorb D-glucose and L-amino acids from the glomerular filtrate (or the primary urine). Stereoselective transport involves a three-point interaction between the substrate and the transporter. Usually one is dealing here with carrier-mediated transport. One type of carrier transport is the so-called facilitated diffusion, which does not require energy and which is unable to operate against a concentration gradient of the substrate involved. The other type is active transport which can pump the substrate against a concentration gradient, but which of course is energy dependent. It appears that the active systems as a rule are the more stereoselective ones. In the living organism many carrier systems are available with a varying degree of selectivity. For instance, several carriers are present for various types of amino acids, such as neutral basic and acidic amino acids. When taken together the carrier systems will give rise to a high degree of stereoselectivity in the distribution of their substrates over the organism.

Now the question arises whether a similar stereoselectivity occurs during the distribution of drugs. Unfortunately little information is available to answer this question. Nevertheless the question seems to be relevant, not only from an academic point of view, which as such already is a legitimate argument, but also from a clinical, therapeutic point of view. Many commonly used drugs in fact are racemic mixtures and it is well-known that their isomers may behave very differently. The differences may come to light in one of the main phases of onset and offset of drug action, viz. drug transport, drug-receptor interaction and drug-biotransformation. Present evidence suggests that stereoselectivity is most pronounced in drug-receptor interaction, that also biotransformation shows substantial stereoselective features but that in case of drug transport stereoselectivity is exception rather than rule. It should be stressed of course that at the subcellular level drug transport is not seldom stereoselective. For instance, the stereoselective uptake and re-uptake of transmitters and drugs in synaptic vesicles or nerve granules is well-documented.

This survey is limited however to macro-distribution effects and in that field our present knowledge is far from comprehensive. Furthermore the data that are available have to be derived from the research of scientists who had totally different objectives when planning their studies, than to establish a possible selective distribution pattern of the isomers of a drug.

Stereoselective distribution of drugs

When transport proceeds by passive diffusion as is usually assumed for drugs, enantiomers will behave similarly, since their physicochemical characteristics are the same. Only when binding to biomacromolecules is involved, the possibility of stereoselective transport exists.

Table 1 shows a compilation of the most plausible mechanisms that may lead to a different distribution behaviour of isomers. Apart from several carrier-mediated transport mechanisms at various sites in the organism it is obvious that also binding to blood and tissue macromolecules may be stereoselective. Binding to plasma proteins usually appears to be not very selective. Erythrocytes on the other hand have binding sites with a high degree of stereoselectivity, for instance with regard to muscarinic cholinergic drugs[1], but most probably there one is dealing with pharmacologic receptors. Erythrocytes further have a high binding capacity for some drugs, notably sulfonamide diuretics[2], but in this case there is no indication of stereoselectivity. A similar situation exists with regard to white blood cells.

In the other tissues a binding capacity is present which can easily be some hundred times higher than in blood. Most of the binding sites however are rather unspecific and of the specific saturable sites only a few percent as a rule is stereoselective. As an example it may be noted that the stereoselective binding of levorphanol in mouse brain accounts for only 2% of the total binding capacity[3]. In the following paragraphs the main causes of selective distribution will be discussed.

Table 1. *Possible stereoselective sites and mechanisms in drug distribution. Substances between brackets are known to be handled in a stereoselective way by the respective mechanisms. References and further details are given in the text.*

A. Carrier-mediated transport	
gastro-intestinal tract	absorption (glucose, amino acids, ascorbic acid)
kidney	tubular secretion active reabsorption (glucose, amino acids)
liver	uptake in hepatocytes (amino acids, tri-iodo-thyronine) secretion into bile
brain	choroid plexus (ascorbic acid) blood-brain barrier (glucose, amino acids, vitamins, hormones, mono-carboxylic acids)
B. Binding to macromolecules	
blood	albumin (tryptophan, benzodiazepines; coumarin derivatives?) globulins (thyroxine, aldosterone, progesterone) blood cells (?)
other tissues	very large binding capacity few stereoselective sites (receptors?)

Carrier-mediated absorption

Intestinal absorption of glucose[4], amino acids[5] and ascorbic acid[6] has been shown to proceed mainly via carrier transport, which at least in part is stereoselective. In the intestinal crypt and villus cells probably two carrier transport mechanisms are operating. The first, at the brush border membrane, is active and stereoselective whereas the second, at the basolateral membrane, has the characteristics of facilitated diffusion without important stereoselectivity. Carrier-mediated absorption obviously is saturable and therefore leads to dose-dependent absorption of the substrate. On basis of the observation that such a saturable absorption is also found for several drugs (a.o. tetracycline, phenytoin), recently a theory was proposed implying that in the absorption of drugs Michaelis-Menten type kinetics not seldom play an important role[7]. Up to now in pharmacokinetics drug absorption as a rule is considered to be a first-order process but often unexplained deviations from first-order kinetics are found. Therefore this topic needs to be studied in much more detail and then also a possible stereoselectivity should be considered, for which no indications are available yet.

Carrier-mediated uptake in blood cells

Also with regard to carrier-mediated uptake in red and white blood cells important information still is lacking. Again glucose and amino acids are taken up in these cells, but the stereoselectivity of this uptake is not established. It is interesting that propranolol isomers inhibit the uptake of glucose in the erythrocyte in a competitive way and that both isomers are equally potent in this respect[8].

In the leucocytes ascorbic acid is taken up by an active transport mechanism, which appears to be inhibited by salicylates[9]. This observation suggests that also the acidic anti-inflammatory analgesics may be taken up by leucocytes. In view of the fact that the leucocyte is at least part of the site of action of anti-inflammatories this is a relevant speculation. Furthermore it would be extremely interesting to know whether such an uptake is stereoselective, since it is well-known that there is a large difference in anti-inflammatory potency of the isomers of for instance ibuprofen and naproxen[10]. Unfortunately again we have no data concerning this intriguing question.

Carrier-mediated transport in the kidney

In the proximal tubular system of the kidney many carrier systems are operating. Roughly these can be divided in two categories: tubular secretion and active reabsorption. Stereoselectivity of this transport has been established, but especially in the active reabsorption of glucose[11,12] and amino acids[13]. Similar to the situation in the small intestine both luminal and antiluminal carriers participate in the reabsorptive mechanism. The transport at the luminal site, located in the brush border membrane, is active and highly selective whereas the antiluminal transport, located in the basolateral membranes of the cells surrounding the proximal tube, appears to be only facilitated diffusion and much less stereoselective. For the amino acids furthermore at each site several types of carriers are available with varying degree of selectivity for the various amino acids. It is not known whether also drugs have affinity for the various amino acid and hexose transporters although it does not seem unlikely a priori that, for instance, glycine conjugates of aromatic acids would be transported by amino acid carriers.

Active tubular secretion has been demonstrated not only for several nutrients and endogenous waste products but also for a large variety of acidic and basic drugs. For the acids the active mechanism is located in the basolateral membrane of the tubular cells and a facilitated diffusion mechanism operates in brush border membrane[14]. For the basic compounds the reverse situation prevails. Structure-transport relationships have not yet been established in an adequate way and there is no indication of stereoselective behaviour at all. In one instance, where possible stereoselectivity was studied, viz. the secretion of mandelic acid derivatives[15], no significant differences between the secretion of the isomers was found.

Carrier-mediated transport in liver

In the liver carrier transport is important at at least two different sites: uptake in the hepatocyte and secretion into bile. Apart from that stereoselective carrier transport of amino acids has been shown in liver mitochondria[16].

The uptake in hepatocytes is a partly stereoselective mechanism for L-amino acids[17] and also for L-tri-iodothyronine[18], but not information is available with regard to drugs, except that there is strong evidence that also these in many instances are taken up by carrier systems. The fact that the uptake mentioned is only in part stereoselective is probably related to the existence of several transport systems with different characteristics (among which also passive diffusion).

Secretion into the bile is a well-known carrier system also for several drugs and metabolites. Up to now no information is available concerning possible stereoselective features of this system, mainly because that aspect has not yet been studied.

Carrier-mediated transport in brain

In the brain two main barriers are of interest for this discussion, viz. the choroid plexus, the boundary between the general circulation and the cerebrospinal fluid, (CSF), and the blood-brain barrier between the circulation and the extracellular fluid of the brain. Carrier transport into and out of the CSF is very well documented. At least in one instance transport into the CSF has been shown to be highly stereoselective, viz. for L-ascorbic acid[19]. Passive diffusion, however, also plays a relatively important role in transport of substances into the CSF. It is interesting furthermore that several drugs appear to be actively secreted from the CSF into the general circulation, e.g. penicillins, aminosalicylic acid[20]. Stereoselectivity here has not been investigated up to now.

In the blood-brain barrier also carrier transport exists. This has been demonstrated to be stereoselective with respect to glucose, amino acids, several vitamins and hormones and also for substances like lactate and pyruvate[21]. As usual hardly any information is available with regard to drugs. There is substantial evidence, however, that the transport of a drug like amphetamine over the blood-brain barrier is carrier-mediated in part, but not stereoselective[22,23].

Stereoselective binding to plasma proteins

The binding of drugs and other substances to the major plasma protein, albumin, as a rule is not stereoselective. There are, however, some remarkable exceptions to this general statement. Both for tryptophan[24] and for oxazepam[25,26] and probably for other benzodiazepines as well, there is a large difference in binding between the isomers. It remains uncertain however whether these differences have important implications with regard to overall distribution and consequently effectiveness of the respective compounds. For several other types of drugs, especially the coumarin anticoagulants, sometimes stereoselectivity to a minor degree in the binding to albumin has been reported[27,28]. Some of the consequences of this will be discussed below.

Whereas binding to albumin usually is characterized by a low affinity and high capacity, the reverse holds true for the binding of hormones and hormone-like substances to the specific steroid binding globulins, such as thyroid hormone binding globulin (TBG), corticosteroid binding globulin (CBG) and sex hormone binding globulin. These binding types combine a high selectivity with a low capacity. Indeed, the binding of thyroxine to TBG has been shown to be stereoselective whereas its binding to albumin, which quantitatively is rather important, is not stereoselective at all[29]. A similar phenomenon was found for aldosterone and progesterone: stereoselective binding to CBG and no selectivity at all in the binding to albumin[30].

Stereoselective binding and the volume of distribution

When binding of isomers of a drug to serum albumin is different one may expect that this in pharmacokinetic studies will be reflected in differences in the volumes of distribution of the isomers.

Assuming that the bound drug distributes over a volume V_1 and the free fraction of drug over a volume V_2 the following equation obviously holds true:

$$V_D.C = V_1C + V_2C_f \qquad \text{(eq.1)}$$

in which V_D = the total apparent volume of distribution
C = the total plasma concentration
C_f = the free plasma concentration

This equation can be rewritten as:

$$V_D = V_1 + V_2 \frac{C_f}{C} \qquad \text{(eq.2)}$$

Now it becomes clear that as long as the free fraction is very small ($C_f << C$), V_D will remain largely determined by V_1 and relatively insensitive to small changes in the free fraction. Such a situation may prevail for instance for warfarin. A small difference in the free fraction of the two isomers of warfarin was found[31], but this did not lead to differences in the volume of distribution of the isomers[32]. At least in man, since in rats in some cases significant differences in volume of distribution have been reported[33,34]. Interestingly, simultaneous administration of phenylbutazone in man led to a stereoselective displacement of warfarin from its binding sites and as a result a significant difference between the distribution volumes of the isomers developed[35]. It should be noted, however, that in case of warfarin differences in distribution of the isomers are very small and for all practical purposes irrelevant.

Another interesting example may be found in the kinetics of the enantiomers of phenprocoumon. Again small differences in binding to albumin exist for the isomers. Here, however, this leads to significant differences in distribution volume in man[36], but not in rat[37]. As an example the relevant pharmacokinetic information on phenprocoumon enantiomers in the rat is listed in Table 2. The volume of distribution does not differ between the isomers, but there appears to be a significant difference in their free fractions in plasma. This in turn causes different liver/plasma concentration ratios and these may be largely responsible for the variation in total body clearance. The conclusion may be that the pharmacokinetic differences between S- and R-phenprocoumon are mainly the result of different distribution patterns, which on the other hand are definitely unable to explain the more relevant pharmacological differences between the isomers.

Table 2. *Pharmacokinetic parameters obtained after administration of phenprocoumon isomers to rats, V_D = volume of distribution, C_f = free plasma concentration, C = total plasma concentration and Cl = total body clearance. Data from Schmidt and Jähnchen*[38]

	V_D (ml/kg)	$\frac{C_f}{C}$ 100%	$\frac{\text{liver conc.}}{\text{plasma conc.}}$	Cl (ml/h. kg)
S(–)	403 ± 24	1.13 ± 0.06	6.9 ± 0.5	22.5 ± 1.1
R(–)	406 ± 20	0.76 ± 0.07	5.2 ± 0.2	15.9 ± 1.0

A third drug for which kinetic differences between isomers have been studied in some detail is the beta-blocker propranolol. In man there appeared to be no important differences in pharmacokinetics of the isomers[39] although the levo-isomer in some studies[40] appeared to have a somewhat smaller volume of distribution than the dextro-isomer, which might be related to differences in protein binding. It is doubtful, however, whether this finding may have any therapeutic importance. For propranolol interesting species differences are found. For instance, in the dog a stereoselective presystemic glucuronidation occurs, which is absent in man[40,41]. Table 3 contains kinetic data obtained after varying doses of racemic or levo-propranolol in the rabbit[42]. Obviously the levo-isomer exhibits a longer half-life of elimination and a larger volume of distribution. Both protein binding and binding to tissue components, including receptors, may be stereoselective to some degree. In this connection sometimes a preferential uptake of the active isomer is suggested. The facts that the volume of distribution as well as the final half-life increase and that the differences between the isomers become smaller with increasing dose are in accordance with that suggestion.

Table 3. *Comparison of the pharmacokinetic behaviour of racemic and l-propranolol in the rabbit after administration of different doses.* $t^{1/2}_{el}$ = *the final half-life of elimination, the other parameters are the same as in Table 2.*
Note that half-life and distribution volume increase with increasing dose. Data from Kawashima and Ishikawa[42]

Dose	0.2 mg/kg		0.4 mg/kg		0.8 mg/kg	
	dl	l	dl	l	dl	l
$t^{1/2}_{el}$ (min)	43	60	52	62	73	85
V_D (1/kg)	8.7	10.4	9.8	12;1	12.7	14.3
Cl (ml/min.kg)	167	151	190	197	155	143

Apparent stereoselective distribution

That apparent stereoselectivity in distribution may arise as a consequence of selective biotransformation or selective excretion is obvious. In such cases a concentration ratio different from unity will arise, without being attributable to differences in the intrinsic distribution kinetics of the enantiomers.

A second factor which has to be taken into account in this respect is the possibility of racemisation or even inversion. For a few drugs it has been found that after administration of separate isomers or of the racemate racemisation or inversion occurred[43,44]. One example is the anti-inflammatory agent ibuprofen, the levo-isomer of which is inverted to the dextro-isomer. As a result of this after administration of the levo-isomer always a smaller apparent volume of distribution is found than after the dextro-isomer. When the inversion process is taken into account however, the volumes of distribution of the isomers appear to be the same.

Conclusions

The conclusion can be summarized shortly as follows:

- stereoselective transport mechanisms are frequently found in the living organism but are directed mainly to nutrients and endogenous substances.
- as far as drug distribution is concerned only in a few instances the possible stereoselectivity has been studied, usually with a negative result.
- it is astonishing to see how many relevant questions with regard to drug

distribution of isomers remain unanswered.

- up to now no indications are found that stereoselective distribution might be an important factor in drug action with pharmacotherapeutic consequences.

References

1. C.R. Mantione and I. Hanin, Mol. Pharmacol. *18*, 28 (1980).
2. H.L.J.M. Fleuren and J.M. van Rossum, J. Pharmacokin. Biopharm. *5*, 359 (1977).
3. A. Goldstein, L.I. Lowney and K.B. Pal, Proc. Nat. Acad. Sci. USA *68*, 1742 (1971).
4. K.-I. Inui, A. Quaroni, L.G. Tillotson and K.J. Isselbacher, Am. J. Physiol. *239* (Cell Physiol. *8*), C190 (1980).
5. E.L. Jervis and D.H. Smyth, J. Physiol. *145*, 57 (1959).
6. G. Toggenburger, M. Landoldt and G. Semenza, FEBS Letters *108*, 473 (1979).
7. J.H. Wood and K.M. Thakker, Eur. J. clin. Pharmacol. *23*, 183 (1982)
8. L. Lačko, B. Wittke and T. Lacko, Arzneim.-Forsch./Drug Res. *29* (II), 1685 (1979).
9. H.S. Loh and C.W.M. Wilson, J. clin. Pharmacol. *15*, 36 (1975).
10. T.Y. Shen, Chem. Eng. News *45*, 10 (1967).
11. R. Kinne, H. Murer, E. Kinne-Saffran, M. Thees and G. Schs, J. Membrane Biol. *21*, 375 (1975).
12. M. Silverman, Can. J. Physiol. Pharmacol. *59*, 209 (1981).
13. S. Silbernagel, Klin. Wochenschr. *57*, 1009 (1979).
14. J.L. Kinsella, P.D. Holohan, N.I. Pessah and C.R. Rose, J. Pharmacol. exp. Ther. *209*, 443 (1979).
15. J.B. Nagwkar, P.M. Patel and M.A. Khambati, J. Pharm. Sci. *62*, 1093 (1973).
16. R.L. Cybulski and R.R. Fisher, Biochemistry *16*, 5116 (1977).
17. H.J. Sips, J.M.M. van Amelsvoort and K. van Dam, Eur. J. Biochem. *105*, 217 (1980).
18. J. Eckel, G.S. Rao, M.L. Rao and H. Breuer, Biochem. J. *182*, 472 (1979).
19. R. Spector and A.V. Lorenzo, Am. J. Physiol. *226*, 1468 (1974).
20. R. Spector and A.V. Lorenzo, J. Pharmacol. exp. Ther. *185*, 642 (1973).
21. W.M. Pardridge and W.H. Oldendorf, J. Neurochem. *28*, 5 (1977).
22. W.M. Pardridge and J.D. Connor, Experientia *29*, 302 (1973).
23. L.R. Steranka, Europ. J. Pharmacol. *76*, 443 (1981).
24. T.P. King and M. Spencer, J. Biol. Chem. *245*, 6134 (1970).
25. W.E. Müller and U. Wollert, Mol. Pharmacol. *11*, 52 (1975).
26. W.E. Müller and U. Wollert, Res. Comm. Chem. Path. Pharmacol. *9*, 413 (1974).
27. K. Veronich, G. White and A. Kapoor, J. Pharm. Sci. *68*, 1515 (1979).
28. E. Schillinger, I. Ehrenberg and K. Lübke, Biochem. Pharmacol. *27*, 651 (1978).
29. S.M. Snyder, R.R. Cavalieri, I.D. Goldfine, S.H. Ingbar and E.C. Jorgensen, J. Biol. Chem. *251*, 6489 (1976).
30. N.V. Aharon and W. Ulick, J. Biol. Chem. *247*, 4939 (1972).
31. A. Yacobi and G. Levy, J. Pharmacokin. Biopharm. *5*, 123 (1977).
32. A. Breckenridge, M. Orme, H. Wesseling, R.J. Lewis and R. Gibbons, Clin. Pharmacol. Ther. *15*, 424 (1974).
33. A. Yacobi and G. Levy, J. Pharmacokin. Biopharm. *2*, 239 (1974).
34. M.J. Fasco and M.J. Cashin, Toxicol. Appl. Pharmacol. *56*, 101 (1980).
35. R.A. O'Reilly, W.F. Trager, C.H. Motley and Howald, J. Clin. Invest. *65*, 746 (1980).
36. E. Jähnchen, T. Meinertz, H.-J. Gilfrich, U. Groth and A. Martini, Clin. Pharmacol. Ther. *20*, 342 (1979).
37. W. Schmidt and E. Jähnchen, J. Pharmacokin. Biopharm. *7*, 643 (1979).
38. W. Schmidt and E. Jähnchen, J. Pharm. Pharmac. *29*, 266 (1977).
39. C.F. George, T. Fenyvest, M.E. Conolly and C.T. Dollery, Europ. J. clin. Pharmacol. *4*, 74 (1972).
40. G.P. Jackman, A.J. McLean, G.L. Jennings and A. Bobik, Clin. Pharmacol. Ther. *30*, 291 (1981).
41. T. Walle and U.K. Walle, Res. Comm. Chem. Path. Pharmacol. *23*, 453 (1979).
42. K. Kawashima and H. Ishikawa, J. Pharmacol. exp. Ther. *213*, 628 (1980).
43. K.J. Kripalani, A. Zein El-Abdin, A.V. Dean and E.C. Schreiber, Xenobiotica *6*, 159 (1976).
44. D.G. Kaiser and G.J. van Giessen, J. Pharm. Sci. *65*, 269 (1976).

Stereoselectivity and Conformation: Flexible and Rigid Compounds

E. Mutschler and G. Lambrecht

Abstract

Since many drug receptors exhibit a high degree of stereo-selectivity towards optical and geometrical isomers, it would not be unexpected for these receptors to possess conformational selectivity as well. Therefore, conformational isomerism in flexible drug molecules may play an important role in determining the interaction of such drugs with their receptors. At the present time, the key to the secret of the 'receptor-active conformation' of flexible molecules might be the development and conformation-activity studies on conformationally constrained rigid or semirigid analogues. Examples are given how this concept has been applied to the analysis of conformation-selectivity relationships in the field of muscarinic agents.

Introduction

One of the principle goals in pharmacology and medicinal chemistry is an adequate description of molecular dynamics of the interactions of pharmacological compounds with their receptors. There are two essential components in such a system which have to be described: the *structure of the drug* and the *structure of the binding sites* of the receptor macromolecule with which the drug interacts. Therefore, any description must be *three-dimensional.* As long as detailed stereochemical information on receptors is lacking, a logical step is to look at the stereochemical anatomy of the drug molecules interacting with a given receptor. A great deal of effort has therefore been devoted to determine the *active conformation* of flexible drug molecules, triggering the response.

One approach to the study of conformational selectivity at drug receptors has been the determination of the *preferred conformation* of the *isolated molecule* (quantum chemical and empirical or force field calculations), of *the molecule in the crystal* (X-ray crystallography), and of *the molecule in solution* (NMR spectroscopy[1–3]. But attempts to deduce the receptor-active conformation of flexible drugs from these studies alone are meaningless. Whether the methodology is theoretical, crystallographic, or spectroscopic, the studied environment (vacuum, crystal, or solution) generally neglects possible conformational changes that can occur in the flexible drug molecule by interaction with the receptor. Such possible conformational changes have often been ignored and focus on the energetically most stable conformer has clouded the issue[4,5].

Attempts to overcome this difficulty resulted in an unique experimental approach by which one is able to determine with a degree of certainty the receptor-active conformation of flexible drugs. This method involves *the development and conformation-activity studies on conformationally constrained rigid or semi-rigid analogues* in which the possibilities of conformational variations are eliminated or greatly reduced.

Methods of reducing conformational flexibility

Four different techniques of controlling the geometry of a flexible drug molecule and of restricting rotations within the molecule may be used:

1. *Making use of steric factors*
   The freedom to rotate may be limited, if the atoms forming a bond have large groups attached to them. In extreme cases, as with some di-ortho-substituted biphenyls, the rotation is sterically so hindered that there are two distinct optical enantiomers[6]. The use of this approach is documented in a series of *diphenhydramines*[7,8].
2. *Making use of bioisosterism*
   There exist only a few studies in the literature in which the *principles of bioisosterism* were used to deduce stereostructure-activity relationships. One example of this approach is provided by pharmacological and stereochemical data comparing *sulfur and selenium congeners of acetylcholine* as nicotinic agonists[9-17].
   Recently we started to control the geometry of pharmacological compounds with the aid of bioisosterism using another approach, namely the substitution of the ammonium group of muscarinic agonists by the sulfonium group. Some of our results are given in the section 'Cyclic Acetylcholine Analogues'.
3. *Making use of multiple bonds*
   The relative positions of atoms attached directly to multiple bonds are fixed. In the case of *double bonds, cis* and *trans isomers* result. In relatively few cases the activities of olefinic cis/trans isomers have been reported. Some examples are the tranquilizing thioxanthenes[18], the anti-depressant zimelidine[19-21], the estrogenic stilbenes[22-25], and cis- and trans-4-aminocrotonic acids as *GABA analogues*[26-28].
4. *Making use of cyclisation*
   The conformational mobility of flexible drugs can be further reduced by incorporation of various parts of the molecules into different ring structures. It is a simple matter to find rigid or semirigid skeletons into which the *essential structure* of drugs (the appropriate functional groups for potential activity) may be incorporated, and much attention has been focused on the receptors of *acetylcholine*[2,4,29-33,72-76], *noradrenaline*[34-36], *dopamine* [37-45,91], *tryptamines*[46-49], *morphinomimetics*[50-62], *histamine*[63-65], *GABA*[27,28,66-71], *neuroleptics*[77,79], *tricyclic antidepressants*[80] *glutamic acid*[81], *amphetamine*[82-86], *amino acids*[87-89], and *hallucinogens*[90], respectively. In the section 'Cyclid Acetylcholine Analogues' examples are given how this concept has been applied in our laboratory to the analysis of stereostructure-activity (-selectivity) relationships in the field of muscarinic agents.

Advantages of controlling the geometry of flexible drugs

The major advantages of reducing conformational flexibility are:

1. The *configuration* of the active pharmacophoric conformation can be determined, and
2. the *key functional groups* are rigidly held in one position, and a semirigid structure constrains these groups to certain limited values.

Within a set of rigid or semirigid analogues, it is assumed that only those which fit the receptor will be active. When stereochemistry alone is responsible for differences in the degree of biological activity between the flexible parent drug and the constrained derivatives, from a comparison of the stereochemistry of the active analogues with possible conformations of the original drug one can make conclusions with respect to the conformation of the parent substance which reacts with the receptor, and the barriers to conformational alteration during drug-receptor interaction.

Disadvantages and limitations of controlling the geometry of flexible drugs

The use of conformationally constrained analogues of flexible drug molecules to investigate the participation of conformational isomerism in drug activity appears to be a feasible approach, although it is not without *pitfalls.* It is difficult (impossible) to control the geometry of a flexible drug without also *changing some other physicochemical properties* of the prototyp drug. To produce a conformationally constrained analogue, new atoms and/or bonds must be added, and this may impart different chemical and physical properties, which must be taken into consideration when interpreting biological data[71-76,98-106].

There is another problem in relating conformational isomerism with biological activity using rigid or semirigid analogues. Rigidity may deny a molecule the opportunity of undergoing a necessary conformational change during its interaction with the receptor. *Too rigid conformers* may, therefore, be sometimes misleading as some degree of flexibility may possibly be needed for interaction with the receptor, if only for sake of mutual adaptability[92].

It has been proposed[93,94] that molecular flexibility finds expression in the *kinetic parameters* of drug receptor interaction. Lass et al.[95], studying the time course of the action of muscarinic antagonists in carp atria, found that the dissociation of rigid antagonists was very much prolonged as compared to flexible drugs of the same affinity. By way of contrast, Wasserman et al.[96] reported that flexibility of structure plays no role in the mechanism of action of acetylcholine at its receptor, using high constrained depolarizing ligands.

The maximum precision of a stereostructure-activity relationship is determined by the *precision of the biological data* used in the study. Therefore, tests which prove that the drugs under investigation act directly via a common receptor are of special importance. In the case of *agonists,* the correlation of potency alone with geometrical factors, neglecting affinity and intrinsic activity of the compounds may sometimes give misleading results[97].

Cyclic acetylcholine analogues

1. Piperidine-, quinuclidine- and thiacyclohexane-derivatives

Conformational isomerism in the acetylcholine molecule may play an important role in determining the interactions with different receptors, such as the nicotinic, the muscarinic, and the acetylcholinesterase receptor surface. Much interest has been focused on the torsion angle of the N-C-C-O fragment (τ_2) (Fig. 1), and especially muscarinic activity has been thoroughly examined[1,2,4,29-33].

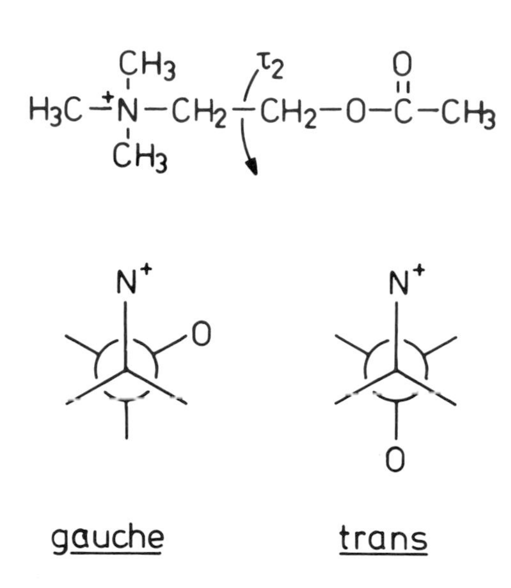

Figure 1. *Structure and conformations of acetylcholine.*

X-ray[10] and *NMR*[1,13,107] studies have shown the N-C-C-O grouping of acetylcholine to exist predominantly in the gauche conformation (Fig. 1). A number of *theoretical, empirical,* and *quantum mechanical calculations*[1,15,108] predict the gauche as the most stable conformer, too. The *global energy minimum* at τ_2 = 60-80 degrees is surrounded by a relatively large zone of low energy, and a number of other *local minima* are present, including one, *13 kJ/mol* above the global minimum, at τ_2 = 180 degrees, corresponding to a trans conformation (Fig. 1)[108]. However, if the free energy of binding of substrates and inhibitors to acetylcholinesterase *(17 kJ/mol)*[109] is taken as a realistic figure for the muscarinic receptor, it is not difficult to rationalize interactions between acetylcholine and the muscarinic receptor that can account for the energy needed to stream over the acetylcholine trans/gauche barrier. Questions arising at this stage are:

1. Does the active conformation of acetylcholine at the muscarinic receptor correspond with its preferred stereochemistry *(= gauche conformation)* or is the energetically less favoured form *(= trans conformation)* required for muscarinic action?
2. If, however, the form active at the muscarinic receptor is the less favoured trans conformation, does the interaction of acetylcholine with the muscarinic receptor lead to the generation of more of the trans conformer up to the point where a new equilibrium is reached? *Is the muscarinic receptor a 'conformerase'?*

To find an answer to the above mentioned questions, about 15 years ago, we started a stereostructure-activity study using Schueler's (R,S)-*3-acetoxy-N-methyl-piperidine methiodide* **I**[101] as a chemical starting point (Fig. 2).

Because it was well known that sometimes tertiary amines are actually more potent as muscarinic agonists than their quaternary derivatives[31,72], and that the muscarinic receptor shows high stereoselectivity[31], we synthesized in the first instance the tertiary (**II**) and quaternary (**I**) individual enantiomers of 3-acetoxy-N-methyl-piperidine (Fig. 2 and 4) and tested these compounds for their muscarinic activity[76,105,111–114].

As shown in Table 1, all the semirigid analogues, based on the piperidine skeleton, are very weak muscarinic agonists. These results, obtained in functional

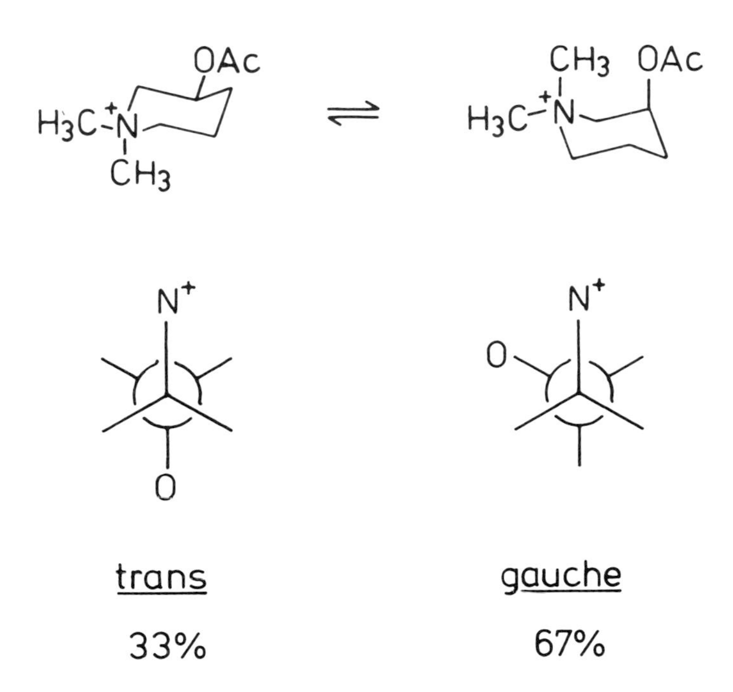

Figure 2. *Chair conformations of (S)-3-acetoxy-N-methyl-piperidine methiodide* **I.** *$\Delta G^\circ = 1.8$ kJ/mol.*

Table 1. *Muscarinic activities of cyclic acetylcholine analogues in the piperidine series*[a].

no.	Compound	pD_2
	Acetylcholine	7.51
(R)-**I**	(R)-3-Acetoxy-N-methyl-piperidine methiodide	3.40[b]
(S)-**I**	(S)-3-Acetoxy-N-methyl-piperidine methiodide	3.92
(R)-**II**	(R)-3-Acetoxy-N-methyl-piperidine	2.99[b]
(S)-**II**	(S)-3-Acetoxy-N-methyl-piperidine	3.68

a Guinea pig isolated left atrium preparation, negative effects on the force of contraction. All experiments were performed in the presence of hexamethonium (0.15 mmol/l), and di-isopropyl fluorophosphate (0.05 mmol/l).
b Partial agonists.

experiments, were confirmed in binding studies[115]. Excluding considerable differences in the electronic structures between acetylcholine[116] and the piperidines[117], and remembering that muscarinic agonists and partial agonists activity is relatively independent of lipophilicity[118], the low muscarinic potency of the semirigid piperidines **I** and **II** might only be seen as a result of structural and stereochemical restrictions. Possibly too much energy is needed for the cyclic analogues to reach the required muscarinic-essential conformation during agonist-receptor interaction, possibly combined with direct steric hindrance caused by the supporting structure of the compounds. But these questions remained, for the time being, to be answered, and it was not possible to draw conclusions concerning the active conformation of acetylcholine.

However, the results of Mashkovsky[119] on the muscarinic activity of racemic *3-acetoxy-quinuclidine* **(III)** and its methiodide **(IV)** suggested an approach to our problem because these esters can be seen as rigid high energy boat form conformers of the semirigid piperidines, and the racemic tertiary ester was found to be a strong muscarinic agonist. Therefore, the individual tertiary and quaternary enantiomers of 3-acetoxy-quinuclidine (Fig. 3) were synthesized and tested for muscarinic activity[76,105,113,120].

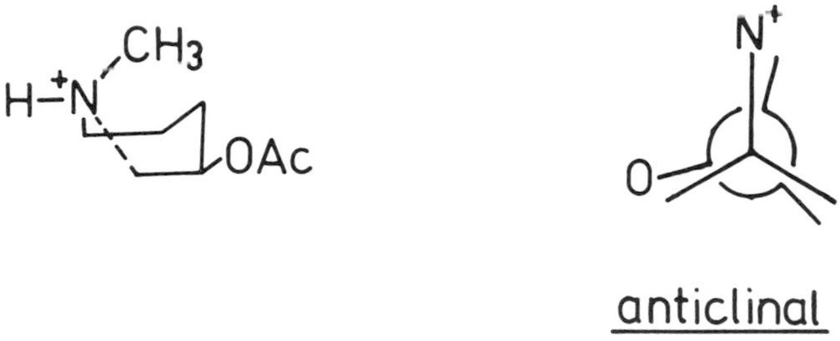

Figure 3. *The structures of the (S)-enantiomers of 3-acetoxy-quinuclidines; R = H* **(III)**, *CH_3***(IV)**[121-123].

As shown in Table 2, the tertiary quinuclidine esters **III** possess remarkable muscarinic potency, and the more potent (S)-enantiomer may be used as a model for discussing fit to the receptor. The muscarinic potency of the corresponding quaternary compounds **IV** is much weaker, probably caused by steric hindrance due to the additional equatorial N-methyl group of the compounds. Binding data for the enantiomers of **III** and **IV** were consistent with their pharmacological activity[124,125].

Table 2. *Muscarinic activities of cyclic acetylcholine analogues in the quinuclidine series*[a]

no.	Compound	pD_2
	Acetylcholine	7.51
(R)-**III**	(R)-3-Acetoxy-quinuclidine	5.02
(S)-**III**	(S)-3-Acetoxy-quinuclidine	6.10
(R)-**IV**	(R)-3-Acetoxy-quinuclidine methiodide	3.97
(S)-**IV**	(S)-3-Acetoxy-quinuclidine methiodide	3.56[b]

a Guinea pig isolated left atrium preparation, negative effects on the force of contraction. All experiments were performed in the presence of hexamethonium (0.15 mmol/l), and di-isopropyl fluorophosphate (0.05 mmol/l).
b Partial agonist.

The differences in *charge distribution* between acetylcholine[116], the piperidines[117], and the quinuclidines[126] are, however, very small, while their muscarinic potencies are far from equal. It is thus not possible to explain the differences in muscarinic potency with differences in the electronic structures.

From a comparison of the muscarinic potency of all the structures in the piperidine (**I** and **II**) and the quinuclidine series (**III** and **IV**) (Table 1 and 2) with their stereochemical properties, and using the strong muscarinic potency of the tertiary ester (S)-**III** for reference, we have postulated that the energetically unfavoured tertiary *cis-(S)-boat* form, shown in Fig. 4, with one axial nitrogen methyl group is a muscarinic-essential conformation of the 3-acetoxy-piperidines. Increase in transition-state energy as these molecules try to adopt the active boat conformation seems to be reflected in the lowered muscarinic potency. The muscarinic receptor-agonist interaction seems not to be capable of overcoming the conformational energy barriers between boat and chair forms of the 6-membered heterocyclic saturated ring[117]. The results of experiments involving both the weakly active enantiomers **IV** indicated, that steric hindrance due to the additional N-methyl group might contribute to the low muscarinic potency of the quaternary piperidine esters **I**.

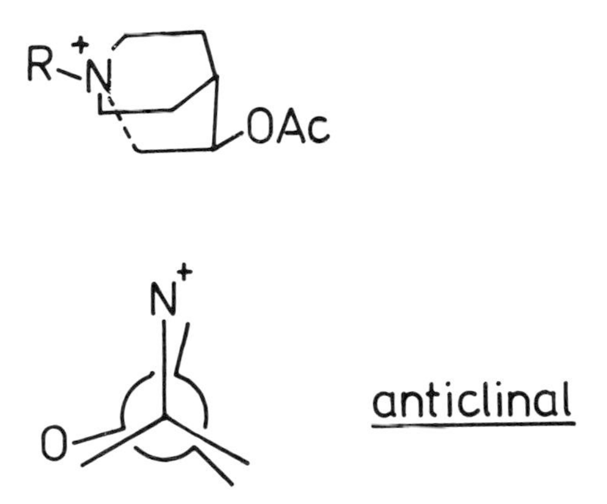

Figure 4. *Potential muscarinic-essential (S)-boat conformation of 3-acetoxy-N-methyl-piperidine* (**II**).

The above mentioned experiments could, however, not exclude the possibility that also an energetically unfavoured chair form of the piperidine analogues with a proper spatial positioning of pharmacophoric groups is a muscarinic-essential conformation. Looking therefore for such a chair conformer, the stereochemistry of the 3-acetoxy-piperidines **I** and **II** was investigated with ^{1}H-NMR spectroscopy[113], and quantum chemical calculations using EHT and CNDO/2 MO methods[117].

With 3-acetoxy-N-methyl-piperidine methiodide (**I**), population analyses assuming an equilibrium system between the two chair conformers shown in Fig. 2 have been carried out. It was found that the chair conformation with the axial acetoxy group (N-C-C-O = gauche) is more stable than the one with the equatorial acetoxy group (N-C-C-O = trans), but the energy difference is rather small ($\Delta G° = 1.8$ kJ/mol).

In the case of the tertiary protonated 3-acetoxy-piperidines **II** four chair conformations should be taken into account. Tertiary piperidinium salts equilibrate according to the scheme in Fig. 5 as a result of two processes: the chair/chair inversion of the ring system and the nitrogen pyramidal inversion via the free amines and reversible deprotonation/protonation. Because of the relatively low value for the nitrogen inversion barrier of about 38 kJ/mol[127], cis/trans isomers can not be isolated, and only two isomers, (R) and (S), exist. The results of the population analyses can be summarized as follows:

The chair conformation with the axial acetoxy group (N-C-C-O = gauche) and the equatorial N-methyl group is the stable one, and the inverse chair with the equatorial acetoxy group (N-C-C-O = trans) and axial N-methyl group is the conformer with the highest energy. The difference in conformational energy between these two conformers is rather high, about 15 kJ/mol.

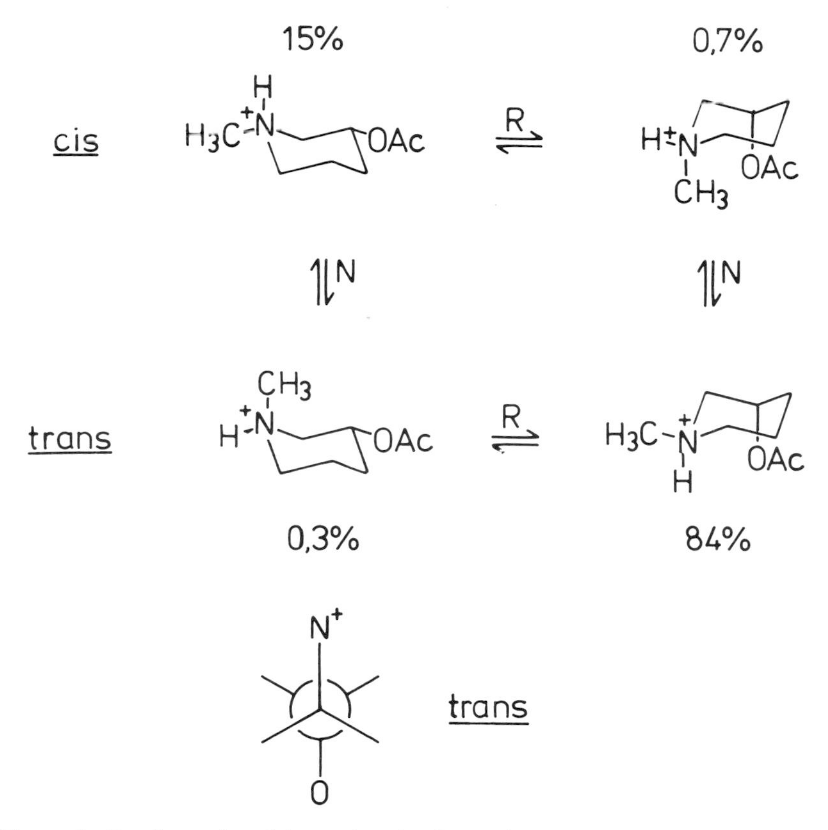

Figure 5. *Conformational isomerism in the tertiary protonated (R)-3-acetoxy-N-methyl-piperidine* (**II**). *R = ring inversion; N = nitrogen pyramidal inversion via the free amines.*

However, comparing the proposed muscarinic essential boat conformation (Fig. 4) and the highest energy chair conformer of the tertiary 3-acetoxy-N-methyl-piperidine (Fig. 5), it is obvious that these two structures have some relevant properties in common:

- the same spatial positioning of the acetoxy and the N-methyl group (equatorial and axial, respectively),
- trans arrangement in the N-C-C-O fragment (anticlinal/antiplanar), and
- both are high energy conformers.

The question here arising was: 'How best to investigate the possibility of participation of the highest energy chair conformer of compound **II** in muscarinic activity?' If this chair form is an active conformation with respect to the muscarinic receptor, then each alteration of the molecule which increases the population of this conformer should increase the muscarinic potency. This can be done by *bioisosteric* *substitution* of the *ammonium group* in the piperidines **I** and **II** by the corresponding *sulfonium group.*

Sulfonium salts like their nitrogen analogues possess a tetrahedral configuration, in which the non-bonded pair of 3s electrons serves as the fourth substituent. The sulfur atom may form a chiral center, and steroisomers can be isolated, since the energy barrier for pyramidal inversion is substantially higher (about 100 kJ/mol) than it is in the case of the corresponding ammonium compounds[111]. Therefore, four stable isomers, cis-(R) and -(S) and trans-(R) and -(S), of the sulfonium analogue of compound **II** may be isolated (Fig. 6).

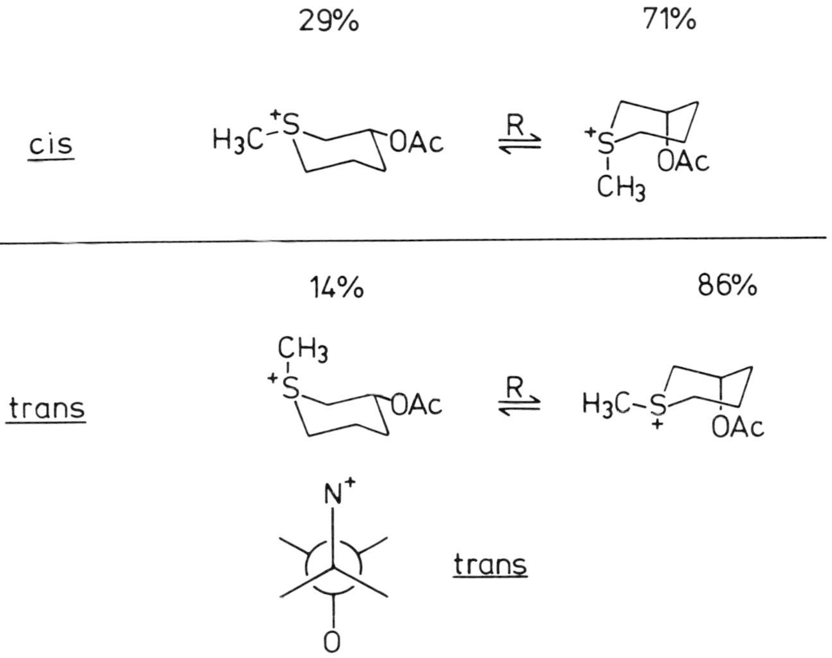

Figure 6. *Conformational isomerism in (1S,3R)-cis-* (**V**) *and (1R,3R)-trans-3-acetoxy-1-methylthiane* (**VI**). *R = ring inversion.*

Quantum chemical[111,117,128] and empirical force field[129] calculations, X-ray crystallography[111,128-130], and NMR spectroscopy[111,128,131-133] have shown that the sulfonium analogues have another important advantage: as a result of the low value of the difference in conformational energy for the sulfur methyl group (0.76 kJ/mol; in the piperidine series, this value amounts to 11-14 kJ/mol), the population of conformers having an axial methyl group in the onium center is much higher than in the piperidine series.

In summary, the replacement of the ammonium group in **II** by the sulfonium group leads to a higher population (by a factor of about 47) of the possibly active chair conformation with axial heteroatom methyl group and equatorial acetoxy group in the trans series of the sulfonium analogues (Fig. 6). Furthermore, both the chair conformers of the sulfonium compounds **V** and **VI** are nearly equivalent in energy, and barriers to ring inversion are low and easily overcome by energy derived from the thermal motion of the molecules or by that released on formation of the agonistreceptor complex itself. If the hypothesis is correct, we should therefore measure a much higher muscarinic potency for one of the enantiomers of the trans-sulfonium compound **VI** than for the corresponding ammonium derivatives **I** and **II**, and the cis-sulfonium ester **V**, respectively. This was in fact the case. The quantitative data for muscarinic activity of the sulfonium analogues are collected in Table 3[111,131,134,135]

Table 3. *Muscarinic activities of cyclic acetylcholine analogues in the thiane series*[a]

no.	Compound	pD_2
	Acetylcholine	7.51
(±)-**V**	(±)-cis-3-Acetoxy-1-methyl-thianium iodide	3.56
(1R,3R)-**VI**	(1R,3R)-trans-3-Acetoxy-1-methyl-thianium perchlorate	6.50
(1S,3S)-**VI**	(1S,3S)-trans-3-Acetoxy-1-methyl-thianium perchlorate	4.85

a Guinea pig isolated left atrium preparation, negative effects on the force of contraction. All experiments were performed in the presence of hexamethonium (0.15 mmol/l), and di-isopropyl fluorophosphate (0.05 mmol/l).

The following immediate observations can be made upon examination of the data in Table 1 and 3: the *(1R,3R)-trans-sulfonium ester* is more than two orders of magnitude more active as a muscarinic agonist than the ammonium compounds and the cis-sulfonium ester, respectively. The enantiomers of the trans-ester **VI** differ substantially in muscarinic potency, and the observed stereospecific index amounts to about 45. Binding data for compounds **V** and **VI** were again consistent with their muscarinic activity in vitro[125].
Excluding considerable differences in the *electronic structures*[117,128] and *molecular dimensions*[111,128–130], and again remembering that muscarinic agonist activity is independent of *lipophilicity,* the differences in muscarinic potency among the cyclic semirigid piperidine and thiane analogues might be seen as the result of *structural and stereochemical restriction.*

As a result of the very low population of boat and twist conformers[117], one of the two chair conformers of the *(1R,3R)-trans-3-acetoxy-1-methyl thiane (***VI***;* Fig. 6) may be expected to represent a muscarinic-essential conformation of this agent. The difference in conformational free energy between the two chair forms amounts to only 4.5 kJ/mol. It thus seems highly possible that the conformer with the lower population of 14% (Fig. 6) can be preferentially bound to the muscarinic receptor. The corresponding conformer of the *tertiary 3-acetoxy-piperidine* ester *(***II***;* Fig. 5) with an axial nitrogen methyl group and an equatorial acetoxy group is of much higher energy. In view of the low muscarinic potency of the piperidine ester **II** it might be suggested, that the active muscarinic-essential conformation is the one with the low population (about 0.3%) and that the energy barrier is so high that it precludes a transformation of the low energy chair form with equatorial nitrogen methyl group and axial acetoxy group into the inverse high energy chair during agonist-receptor interaction.

The difference in conformational free energy for both chair conformers of the *quaternary piperidine enantiomers* **I** (Fig. 2) is as small as for the trans-sulfonium analogue **VI**, and the population of the conformer with an equatorial acetoxy group is rather high. The low muscarinic potency of compounds **I** may be seen as a result of direct steric hindrance due to the additional nitrogen methyl group of the compounds.

The transformation of the *cis-sulfonium analogue* **V** (Fig. 6) from any possible conformation into exactly the correct active chair conformation with axial sulfur methyl group and equatorial acetoxy group is not possible without cleaving bonds. On the other hand, this sulfonium ester **V** can exist in the active boat conformation, shown in Fig. 4. But this boat form should be inaccessible during agonist-receptor interaction to some of the conformational modes of the molecule, due to the magnitude of the energy barrier separating the boat and chair conformations[117]. These may be the reasons for the relatively low muscarinic potency of compound **V**.

The structural and stereochemical demands for high muscarinic potency of compounds like the semirigid piperidine (**I** and **II**) and thiane (**V** and **VI**) esters (Fig. 2,5 and 6) can be summarized as follows (Fig. 7):

For drugs of this type, a muscarinic-essential conformation has to be adopted by the agonist molecules at some stage during productive interaction at the muscarinic receptor: a *cis-(S)-boat* or a *trans-(1R,3R)-chair* with only one methyl group in the onium center in the axial orientation and the acetoxy group equatorial (N-C-C-O = anticlinal/antiplanar). This implies that the potency of these molecules will be impaired by structural configurational and/or conformational restrictions unless they can adopt one of the two active conformations, shown in Fig. 7.

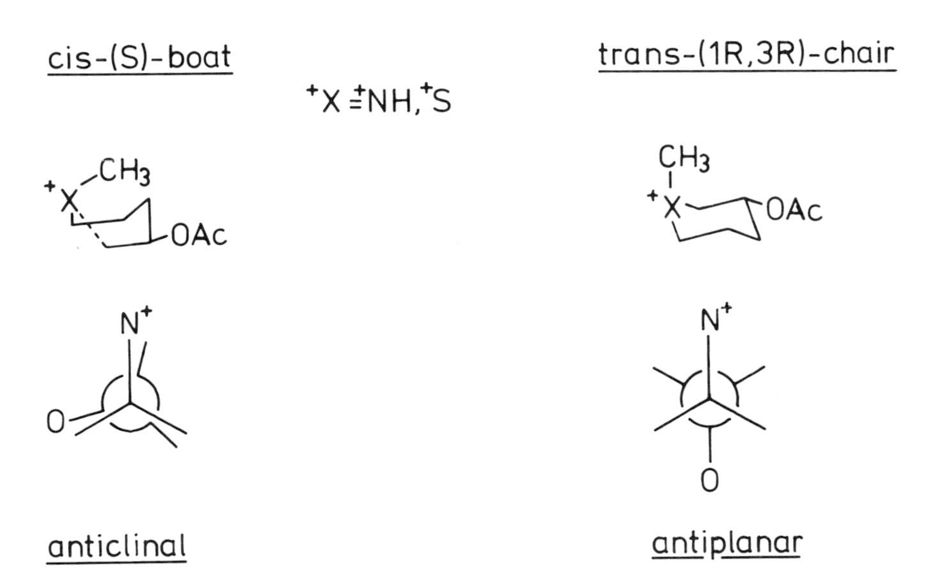

Figure 7. *Proposed muscarinic-essential conformations of piperidine and thiane analogues of acetylcholine.*

Finally, returning to the original questions, it remains to be discussed, whether a gauche or a trans conformation of *acetylcholine* (Fig. 1) is the active muscarinic-essential conformation, and to what extent the muscarinic receptor can create a preferred conformer of acetylcholine when only a few molecules of this are presented by the dissolved drug.

1. Our studies appear to support the idea that a *transoid anticlinal/antiplanar N-C-C-O orientation* is involved in the interaction of acetylcholine with muscarinic receptors even though the preferred solution conformation for this moiety in

acetylcholine is gauche. This is an agreement with a number of other studies, which concur in identifying anticlinal/antiplanar N^+/O conformers as the active muscarinic species, and clearly exclude synclinal gauche forms[2,4,30,32].

2. If the muscarinic receptor is not able to generate more of the trans-N-C-C-O conformer during acetylcholine – receptor interaction, the apparent potency of acetylcholine will simply be the product of the potency for the correct trans conformer and the mole fraction of this conformer in solution. Using a value of about 13 kJ/mol for the energy difference between the gauche and trans conformation of acetylcholine[108], one calculates a population of the trans conformer of less than 1%. That would mean that the potency of acetylcholine was reduced to less than 1% of that had it existed primarily in the trans conformation in solution, and the potency of rigid or semirigid analogues, having a transoid N-C-C-O orientation, should be at least two orders of magnitude higher than the apparent potency of acetylcholine. Rigid or semirigid analogues with such a high muscarinic potency have not yet been reported, due presumably to the fact that a conformational change in the acetylcholine molecule is induced by interaction with the muscarinic receptor, leading to preferential binding of the drug in a trans conformation. In this case, the muscarinic receptor functions as a '*conformerase*'.

2. *Arecoline, isoarecoline, and their sulfonium analogues*

In order to determine the structural and stereochemical requirements around the cationic head for optimum activity at muscarinic sites the cholinergic properties of the semirigid compounds **VII-XII** (Table 4) and their conformational behaviour (quantum chemical calculations using EHT and CNDO/2 MO methods) were investigated[140-142].

Table 4. *Muscarinic activities of cyclic acetylcholine analogues in the arecoline and isoarecoline series*[a]

Structure	no.	Compound	Position of ester group/X	pD_2
		Acetylcholine		7.31
$COOCH_3$; ring with $\overset{+}{X}$–CH_3	**VII**	Arecoline	C-3/NH	6.47
	VIII	Arecoline methiodide	C-3/NCH_3	5.21
	IX	Sulfoarecoline	C-3/S	6.21
	X	Isoarecoline	C-4/NH	5.03
	XI	Isoarecoline methiodide	C-4/NCH_3	6.14
	XII	Sulfoisoarecoline	C-4/S	6.13

[a] Guinea pig isolated left atrium preparation, negative effects on the force of contraction. All experiments were performed in the presence of hexamethonium (0.15 mmol/l), and diisopropyl fluorophosphate (0.05 mmol/l).

Compounds **VII-XII** are directly acting full muscarinic agonists. As shown in Table 4, the muscarinic potency of esters **VII, IX, XI** and **XII** is about one order of magnitude lower than that of acetylcholine, and about ten times higher than that of compounds **VIII** and **X**.

In the tertiary ammonium compounds **VII** and **X**, the half chair conformation with an equatorial methyl group is the most stable one (Fig. 8; Δ E = -33.5 and -32.6 kJ/mol, respectively). The lowest energy conformer of the sulfonium analogues **IX** and **XII** carries the methyl group in the axial orientation (Fig. 8; Δ E = +25.5 and +15.8 kJ/mol, respectively).

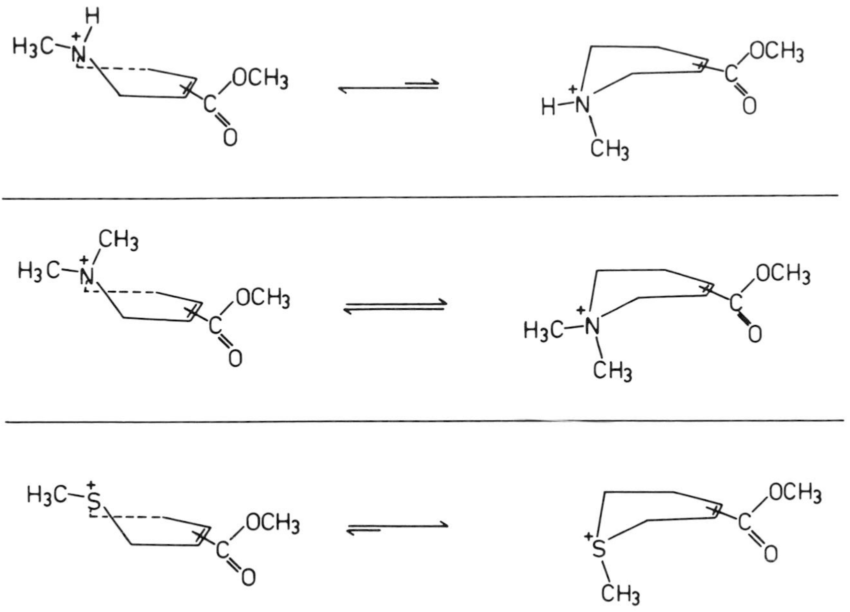

Figure 8. *Structures and conformational equilibria of compounds* **VII-XII.**

On the basis of pharmacological tests and conformational analyses performed with compounds **VII-XII**, it is possible to draw conclusions concerning the structural and stereochemical requirements for activity at muscarinic receptors:

For high muscarinic potency, one methyl group in the onium center is sufficient. In the isoarecaidine series (**X-XII**), a second methyl group impairs the muscarinic potency only slightly, but in the case of the quaternary arecoline **VIII** this second methyl group gives rise to a pronounced steric hindrance. In the active conformation of isoarecoline **X** and sulfo-isoarecoline **XII**, the essential heteroatom methyl group should be in the *axial* position.

In order to further reduce the conformational flexibility of arecaidine and isoarecaidine esters, we have prepared the rigid bridged analogues, shown in Fig. 9, in

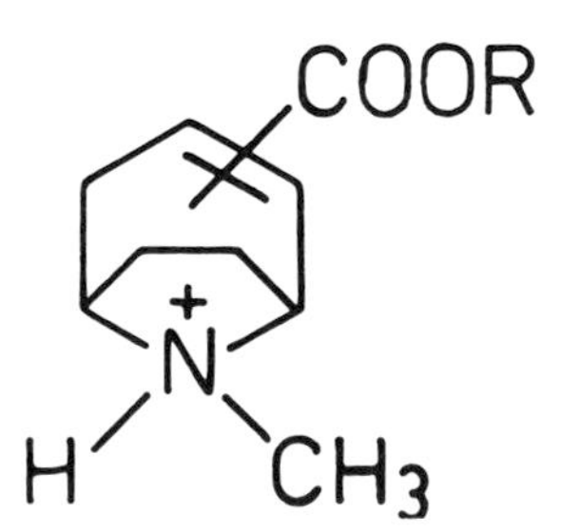

Figure 9. *Conformationally rigid esters which act selectively on muscarinic receptor subtypes. R = alkyl, alkenyl, alkynyl.*

which the tetrahydropyridine ring is forced into one single half-chair-like conformation. Some of these esters act selectively as muscarinic agonists at atrial m_2 receptors [138,139]

Conclusions

Conformationally rigid or semirigid analogues of flexible drugs do offer a powerfull aid to establish the conformational (configurational) requirements of pharmacological receptors, which in turn may lead to enhanced *selectivity* of current drugs, and to design of new ones. Both the conformers, shown in Fig. 7, act selectively as muscarinic agonists. They are feeble as nicotinic agonists, and only poor substrates for acetylcholinesterase[113,124–137]. In addition, the different muscarinic potencies and affinities of the esters (S)-**I** (Table 1) and **V** (Table 3) at atrial and ileal muscarinic receptors[111,113,135] support the view that these receptors can be classified into m_1 in the ileum and m_2 in the atria[139]. This concept of heterogeneity of muscarinic receptors was significantly strenghtened by the discovery of certain rigid esters, shown in Fig. 9, which act selectively as muscarinic agonists at atrial m_2 receptors[138].

Acknowledgement

Our own work, presented in this review, was partially supported by the Deutsche Forschungsgemeinschaft und Dr. Robert-Pfleger-Stiftung. It is a pleasure to acknowledge the close collaboration of Dr. Hammer, Biochemical Department, Istituto De Angeli, Milan (Italy), Prof. Höltje, Institute of Pharmaceutical Chemistry, University of Frankfurt (FRG), Dr. Jensen, Royal Danish School of Pharmacy, Copenhagen (Denmark), and Prof. Triggle, Department of Biochemical Pharmacology, State University of New York at Buffalo (USA). The authors are greatly indebted to Mrs. Ch. Röttger and Mrs. M. Wagner for their valuable assistance in carrying out the pharmacological investigations, and for their help in preparing the manuscript, and providing the figures.

References

1. E.D. Bergmann and B. Pullman, eds., 'Molecular and Quantum Pharmacology', D. Reidel Publishing Company, Dordrecht-Holland, 1974.
2. A.F. Casy, In 'Progress in Medicinal Chemistry', Vol. 11, pp. 1-65, (G.P. Ellis and G.B. West, eds.), North-Holland Publishing Company, Amsterdam, 1975.
3. E.D. Bermann and B. Pullman, eds., 'Conformation of Biological Molecules and Polymers', The Israel Academy of Sciences and Humanities, Jerusalem, 1973.
4. P.S. Portoghese, Ann. Rev. Pharmacol. *10*, 51 (1970).
5. K.W. Reed, W.J. Murray, E.B. Roche and L.N. Domelsmith, Gen. Pharmacol. *12*, 177 (1981).
6. H. Musso and W. Steckelberg, Justus Liebigs Ann. Chem. *693*, 187 (1966).
7. A.F. Harms and W.T. Nauta, J. Med. Chem. *2*, 57 (1960).
8. E.J. Ariens, In 'Biological Activity and Chemical Structure', pp. 1-25 (J.A. Keverling Buisman, ed.), Elsevier, Amsterdam, 1977.
9. G.D. Webb and H.G. Mautner, Biochem. Pharmacol. *15*, 2105 (1966).
10. S. Jagner and B. Jensen, Acta Cryst. B. *33*, 2757 (1977).
11. A. Makriyannis, R.F. Sullivan and H.G. Mautner, Proc. Nat. Acad. Sci. USA *69*, 3416 (1972).
12. H.G Mautner, D.D. Dexter and B.W. Low, Nature New Biology *238*, 87 (1972).
13. P. Partington, J. Feeney and A.S.V. Burgen, Mol. Pharmacol. *8*, 269 (1972).
14. G. Chidichimo, F. Lelj and N. Russo, J. theor. Biol. *66*, 811 (1977).
15. B. Pullman and P. Courriere, Mol. Pharmacol. *8*, 371 (1972).
16. H.S. Aldrich, Int. J. Quantum Chem.: Quantum Biology Symp. *2*, 271 (1975).
17. H.G. Mautner, In 'Molecular and Quantum Pharmacology', pp. 119-129, (E.D. Bergmann and B. Pullman, eds.), D. Reidel Publishing Company, Dordrecht-Holland, 1974.

18. J.D. Dunitz, H. Eser and P. Strickler, Helv. Chim. Acta *47*, 1897 (1964).
19. D. Brown, D.H.T. Scott, D.B. Scott, M. Meyer, D. Westerlund and J. Lundström, Eur. J. Clin. Pharmacol. *17*, 111 (1980).
20. A.R. Coppen, R. Rao, C. Swade and K. Wood, Psychopharmacol. *63*, 199 (1979).
21. S.B. Ross and A.L. Renyi, Neuropharmacol. *16*, 57 (1977).
22. E.C. Dodds, L. Goldberg, W. Lawson and R. Robinson, Nature *141*, 247 (1938).
23. U.V. Solmssen, Chem. Rev. *37*, 481 (1945).
24. E. Walton and G. Brownlee, Nature *151*, 305 (1943).
25. F. von Wessely, Angew. Chem. *53*, 197 (1940).
26. G.A.R. Johnston, D.R. Curtis, P.M. Beart, C.J.A. Game, R.M. McCulloch and B. Twitchin, J. Neurochem. *24*, 157 (1975).
27. P. Krogsgard-Larsen, J. Scheel-Früger and H. Kofod, 'GABA Neurotransmitters', Munksgaard, Copenhagen, 1978.
28. R.D. Allan, D.R. Curtis, P.M. Headley, G.A.R. Johnston, D. Lodge and B. Twitchin, J. Neurochem. *34*, 652 (1980).
29. F. Gualtieri, P. Angeli, M. Giannella, C. Melchiorre and M. Pigini, In 'Recent Advances in Receptor-Chemistry', pp. 267-279, (F. Gualtieri, M. Giannella and C. Melchiorre, eds.), Elsevier, Amsterdam, 1979.
30. G. Lambrecht, Pharmazie *31*, 209 (1976).
31. D.J. Triggle, In 'Chemical Pharmacology of the Synapse', pp. 233-430, (D.J. Triggle and C.R. Triggle, eds.), Academic Press, London-New York-San Francisco, 1976.
32. G. Lambrecht, 'Cyclische Acetylcholine-Analoga', H. und P. Lang, Bern-Frankfurt, 1971.
33. G. Lambrecht and E. Mutschler, In 'Medicinal Chemistry Advances', pp. 117-129, (F.G. de las Heras and S. Vega, eds.), Pergamon Press, Oxford, 1981.
34. H.L. Komiskey, F.L. Hsu, F.J. Bossart, J.W. Fowble, D.D. Miller and P.N. Patil, Eur.
35. L. Bausi, J.M. Khanna and N. Anand, J. Med. Chem. *15*, 23 (1972).
36. E.E. Smissman and R.T. Borchardt, J. Med. Chem. *14*, 377 (1971).
37. P. Burn, P.A. Crooks, F. Heatley, B. Costall, R.J. Naylor and V. Nohria, J. Med. Chem. *25*, 363 (1982).
38. B. Costall, S.K. Lim, R.J. Naylor and J.G. Cannon, J. Pharm. Pharmacol. *34*, 246 (1982).
39. S.-J. Law, J.M. Morgan, L.W. Masten, R.F. Borne, G.W. Arana, N.S. Kula and R.J. Baldessarini, J. Med. Chem. *25*, 213 (1982).
40. A. Davis, J.A. Poat and G.N. Woodruff, Eur. J. Pharmacol. *63*, 237 (1980).
41. H.L. Komiskey, J.F. Bossart, D.D. Miller and P.N. Patil, Proc. Nat. Acad. Sci. USA *75*, 2641 (1978).
42. U. Hacksell, U. Svensson, J.L.G. Nilsson, S. Hjorth, A. Carlsson, H. Wikström, P. Lindberg and D. Sanchez, J. Med. Chem. *22*, 1469 (1979).
43. P.W. Erhardt, R.J. Gorczyuski and W.G. Anderson, J. Med. Chem. *22*, 907 (1979).
44. J.L. Tedesco, P. Seeman and J.D. McDermed, Mol. Pharmacol. 16, 369 (1979).
45. A.S. Horn and J.R. Rodgers, J. Pharm. Pharmacol. *32*, 521 (1980).
46. E. Friedman, E. Meller and M. Hallock, J. Neurochem. *36*, 931 (1981).
47. G.E.A. Coombes and D.J. Harvey, J. Chem. Soc. C 325 (1970).
48. E. Friderichs, W. Back and E. Mutschler, Arch. Pharm. (Weinheim) *308*, 209 (1975).
49. E. Friderichs, W. Back and E. Mutschler, Arch. Pharm. (Weinheim) *308*, 663 (1975).
50. W. Klein, W. Back and E. Mutschler, Arch. Pharm. (Weinheim) *307*, 360 (1974).
51. W. Klein, W. Back and E. Mutschler, Arch. Pharm. (Weinheim) *308*, 910 (1975).
52. T.N. Riley and J.R. Bagley, J. Med. Chem. *22*, 1167 (1979).
53. J.G. Berger, F. Davidson and G.E. Langford, J. Med. Chem. *20*, 600 (1977).
54. S.J. Law, D.H. Lewis and R.F. Borne, J. Heterocycl. Chem. *15*, 273 (1978).
55. A.F. Casy, In 'Progress in Medicinal Chemistry', Vol. 7, pp. 229-284, (G.P. Ellis and G.B. West, eds.), Butterworth, London, 1970.
56. A.F. Casy, In 'Drug Research', Vol. 22, pp. 149-227, (E. Jucker, ed.), Birkhäuser Verlag, Basel, 1978.
57. B.E. Maryanoff, D.F. McComsey, R.J. Taylor and J.F. Gardocki, J. Med. Chem. *24*, 79 (1981).
58. R.F. Borne, S.J. Law, J.C. Kapeghian and L.W. Masten, J. Pharm. Sci. *69*, 1104 (1980).
59. M.E. Rodgers, D.S. Wilkinson, J.R. Thweatt and S.P. Halenda, J. Med. Chem. *23*, 688 (1980).

60. P.A. Crooks and R. Szyndler, J. Med. Chem. *23*, 679 (1980).
61. P.S. Portoghese, A.A. Mikhail and H.J. Kupferberg, J. Med. Chem. *11*, 219 (1968).
62. P.S. Portoghese, J. Pharm. Sci. *55*, 865 (1966).
63. A. Burger, M. Bernabe and P.W. Collins, J. Med. Chem. *13*, 33 (1970).
64. W. Schunack, Arch Pharm. (Weinheim) *306*, 934 (1973).
65. S. Schwarz and W. Schunack, Arch. Pharm. (Weinheim) *312*, 933 (1979).
66. P. Krogsgaard-Larsen, J. Med. Chem. *24*, 1377 (1981).
67. A.J. Porsius, G. Lambrecht, U. Moser and E. Mutschler, Eur. J. Pharmacol. *77*, 49 (1982).
68. D.T. Witiak, K. Tomita, R.J. Patch and S.J. Enna, J. Med. Chem. *24*, 788 (1981).
69. R.M. Black, J.C.S. Perkin I 73 (1982).
70. J.P. O'Donnell, D.A. Johnson and A.J. Azzaro, J. Med. Chem. *23,* 1142 (1980).
71. R.D. Allan, D.R. Curtis, P.M. Headley, G.A.R. Johnston, S.M.E. Kennedy, D. Lodge and B. Twitchin, Neurochem. Res. *5,* 393 (1980).
72. H. Gloge, H. Lüllmann and E. Mutschler, Brit. J. Pharmacol. *27*, 185 (1966).
73. K. Hultzsch, U. Moser, W. Back and E. Mutschler, Arzneim.-Forsch. *21*, 1979 (1971).
74. G. Lambrecht and E. Mutschler, Arzneim.-Forsch. *23*, 1427 (1973).
75. E. Mutschler and K. Hultzsch, Arzneim.-Forsch. *23*, 732 (1973).
76. G. Lambrecht and E. Mutschler, Arzneim.-Forsch. *24*, 1725 (1974).
77. C.J. Crol, D. Dijkstra, W. Schunselaar, B.H.C. Westerink and A.R. Martin, J. Med. Chem. *25*, 5 (1982).
78. J.P. Li and J.H. Biel, J. Med. Chem. *12*, 917 (1969).
79. A.R. Martin, S.H. Kim, H.I. Yamamura and A.S. Horn, J. Med. Chem. *23*, 938 (1980).
80. A.R. Martin, V.M. Paradkar, G.W. Peng, R.C. Speth, H.I. Yamamura and A.S. Horn, J. Med. Chem. *23*, 865 (1980).
81. E.G. McGeer, J.W. Olney and P.L. McGeer, 'Kainic Acid as a Tool in Neurobiology', Raven Press, New York, 1978.
82. B.A. Hathaway, D.E. Nichols, M.B. Nichols and G.K.W. Yim, J. Med. Chem. *25*, 535 (1982).
83. G.L. Grunewald, R.T. Borchardt, M.F. Rafferty and P. Krass, Mol. Pharmacol. *20,* 377 (1981).
84. L.E. Wood, R. Daniels, L. Bauer and J.E. Gearien, J. Pharmac. Sci. *70*, 199 (1981).
85. G.L. Grunewald, T.J. Reitz, A. Hallett, C.O. Rutledge, S. Vollmer, J.M. Archuleta and J.A. Ruth, J. Med. Chem. *23*, 614 (1980).
86. R.I. Mrongovius., A.G. Bolt and R.O. Hellyer, Clin. Exper. Pharmacol. Physiol. *5*, 635 (1978).
87. G.L. Grunewald, S.H. Kuttab, M.A. Pleiss, J.B. Mangold and P. Soine, J. Med. Chem. *23,* 754 (1980).
88. J.R. Sufrin, A.W. Coulter and P. Talalay, Mol. Pharmacol. *15*, 661 (1979).
89. M. Bernabé, O. Cueavas and E. Fernandez-Alvares, Eur. J. Med. Chem. *14*, 33 (1979).
90. D.E. Nichols, R. Woodarol, B.H. Hathaway, M.T. Lowy and G.K.W. Yim, J. Med. Chem. *22*, 458 (1979).
91. J.L. Neumeyer, G.W. Arana, V.J. Ram, N.S. Kula and R.J. Baldessarini, J. Med. Chem. *25*, 990 (1982).
92. B. Pullman, In 'Molecular and Quantum Pharmacology', pp. 9-36. (E. Bergmann and B. Pullman, eds.), D. Reidel Publishing Company, Dordrecht-Holland, 1974.
93. A.S.V. Burgen, G.C. K. Roberts and J. Feeney, Nature *253*, 753 (1975).
94. W.H. Beers and E. Reich, Nature *228*, 917 (1970).
95. Y. Lass, S. Akselrod, B. Gavish, S. Cohen and A. Fisher, Experientia *35*, 650 (1979).
96. N.H. Wassermann, E. Bartels and B.F. Erlanger, Proc. Nat. Acad. Sci. USA *76*, 256 (1979).
97. E. Shefter, In 'Cholinergic Ligand Interactions', pp. 83-117. (D.J. Triggle, J.F. Moran and E.A. Barnard, eds.), Academic Press, New York, 1971.
98. B.C. Barrass, R.W. Brimblecombe, D.C. Parkes and P. Rich, Brit. J. Pharmacol. *34,* 345 (1968).
99. E. Mutschler, H. Scherf and O. Wassermann, Arzneim.-Forsch. *17,* 833 (1967).
100. G. Lambrecht and E. Mutschler, Arch. Pharm. (Weinheim) *308*, 455 (1975).
101. G. Aksnes and P. Frøyen, Acta Chem. Scand. *20*, 1451 (1966).
102. J. Dolby, K.-H. Hasselgren, J.L.G. Nilsson and M. Elander, Acta Pharm. Suecica *8,* 97 (1971).

103. M. Freifelder, J. Org. Chem. *29*, 2895 (1964).
104. E. Wenkert, K.G. Dave, F. Haglid, R.G. Lewis, T. Oishi, R.V. Stevens and M. Terashima, J. Org. Chem. *33*, 747 (1968).
105. G. Lambrecht and E. Mutschler, In 'Molecular and Quantum Pharmacology', pp. 179-187. (E. Bergmann and B. Pullman, eds.), D. Reidel Publishing Company, Dordrecht-Holland, 1974.
106. C.A. Grob, A. Kaiser and E. Renk, Chem. Ind. 598 (1957).
107. A. Makriyannis, J.M. Theard and H.G. Mautner, Biochem. Pharmacol. *28*, 1911 (1979).
108. B. Pullman and P. Courrière, Mol. Pharmacol. *6*, 612 (1972).
109. B. Belleau and J.L. Lavoie, Can. J. Biochem. *46*, 1397 (1968).
110. F.W. Schueler, J. Am. Pharm. Ass. *45*, 197 (1956).
111. H.-D. Höltje, B. Jensen and G. Lambrecht, In 'Recent Advances in Receptor Chemistry', pp. 281-302. (F. Gualtieri, M. Giannella and C. Melchiorre, eds.), Elsevier, Amsterdam, 1979.
112. G. Lambrecht, Experientia *32*, 365 (1976).
113. G. Lambrecht, Eur. J. Med. Chem. *11*, 461 (1976).
114. G. Lambrecht, Arch. Pharm. (Weinheim) *313*, 368 (1980).
115. R. Hammer, private communication.
116. B. Pullman and P. Courriere, Theoret. Chim. Acta *31*, 19 (1973).
117. H.-D. Höltje, Arch. Pharm. (Weinheim) *311*, 311 (1978).
118. K.J. Chang, R.C. Deth and D.J. Triggle, J. Med. Chem. *15*, 243 (1972).
119. M.D. Mashkovsky, In 'Proceedings of the 1st International Pharmacology Meeting', Vol. 7, pp. 359-366, (K.J. Brunings and P. Lindgren, eds.), Pergamon Press, London, 1963.
120. G. Lambrecht, Arch. Pharm. (Weinheim) *309*, 235 (1976).
121. R.W. Baker and P.J. Pauling, J.C.S. Perkin II 2340 (1972).
122. B. Belleau and P.J. Pauling, J. Med. Chem. *13*, 737 (1970).
123. A. Meyerhöffer, J. Med. Chem. *15*, 994 (1972).
124. B. Ringdahl, F.J. Ehler and D.J. Jenden, Mol. Pharmacol. *21*, 594 (1982).
125. D.J. Triggle, private communication.
126. H. Weinstein, S. Maayani, S. Srebrenik, S. Cohen and M. Sokolovsky, Mol. Pharmacol. *11*, 671 (1975).
127. F.G. Ridell and H. Labaziewicz, J.C.S. Chem. Commun. 766 (1975).
128. H.-D. Höltje, B. Jensen and G. Lambrecht, Eur. J. Med. Chem. *13*, 453 (1978).
129. B. Jensen, Acta Chem. Scand. B *35*, 607 (1981).
130. B. Jensen, Acta Chem. Scand. B *33*, 359 (1979).
131. G. Lambrecht, Eur. J. Med. Chem. *12*, 41 (1977).
132. G. Lambrecht, Arch. Pharm. (Weinheim) *310*, 1015 (1977).
133. G. Lambrecht, Arch. Pharm. (Weinheim) *311*, 636 (1978).
134. G. Lambrecht, Arch. Pharm. (Weinheim) *313*, 368 (1980).
135. G. Lambrecht, Arzneim.-Forsch. *31*, 634 (1981).
136. G. Lambrecht, Arch. Pharm. (Weinheim) *315*, 646 (1982).
137. R.B. Barlow and A.F. Casy, Mol. Pharmacol. *11*, 690 (1975).
138. G. Höfling and E. Mutschler, unpublished results.
139. R.B. Barlow, K.N. Burston and A. Vis, Brit. J. Pharmacol. *68*, 141 P (1980).
140. H.-D. Höltje, G. Lambrecht, U. Moser and E. Mutschler, Arzneim.-Forsch., in press.
141. E. Mutschler, H.-D. Höltje, G. Lambrecht and U. Moser, Arzneim.-Forsch., in press.
142. U. Moser, G. Lambrecht, E. Mutschler and J. Sombroek, Arch. Pharm (Weinheim), in press.

The Stereochemistry of Binding to Receptors

A.S.V. Burgen

Abstract

Examination of stereoisomeric pairs bound to receptors by n.m.r. shows that part of the ligand may bind in the same way in both isomers if not prevented by steric factors, other parts of the ligand may be prevented from binding at all and may remain essentially surrounded by solvent. Similar finding may occur if the binding of the same fragment is changed by chemical substitution. Where steric factors make it impossible for any fragment of the ligand to bind in the base orientation, global adjustments occur. Substitutions may also lead to major reorientations of binding with increased affinity.

It is a truism that the activity of drugs depends on the formation complexes with receptors and that the stability of these complexes is dependent on the force field generated by the apposition of atomic groupings in the two partners.

Since receptors are macromolecular with a more or less well defined three dimensional structure, the contact with the drug will also be three dimensional, and within this frame of reference chiral molecules are *chemically* distinct. Indeed it is well known that chirality is imposed on non-chiral molecules when they bind to receptors. It is therefore not unexpected that optical enantiomers will usually have different binding affinities for a receptor and of course drug receptors are not unique in this regard.

It has been frequently proposed and widely accepted that a good test of 'specific' binding to a receptor is obtained by using the binding of the less favoured ('inactive') enantiomer as a measure of non-specific binding. Since it is a different structure there is no sound basis for this and it is less correct than self-saturation or cross saturation used for the same purpose. It should be noted that selectivity between enantiomers has been found in non-biological systems, a striking example being the binding of opiates to glass fibre filters[1].

Simple binding data have distinct limitations, for instance, when subtypes of receptor are present it is not always straightforward to relate binding to any one of them to a particular pharmacological activity. Secondly, binding does not in itself distinguish between the different pharmacology of agonists and antagonists, although subsidiary features such as sensitivity to guanine nucleotides or ions may be indicative. Despite these snags, in general, binding differences between enantiomeric pairs do correlate reasonably well with the pharmacological data.

Since enantiomeric pairs are good probes for the detail of a receptor surface, it is interesting to see how they discriminate between receptor subtypes such as those of the muscarinic receptor[2] (Table 1). The discrimination between enantiomers is reduced in the lower affinity forms of the receptor, nevertheless if the discrimination is related to the mean affinity it actually improves as one proceeds to the lower affinity form. Note that for all the subtypes both enantiomers bind less well than to the 'parent' compound acetylcholine.

The reaction of the mercurial PCMB with sulphydryl groups in the muscarinic re-

Table 1. *Muscarinic receptor sub-types. Binding constants (M^{-1})*

	SH	H	L
Acetylcholine	5×10^7	3.2×10^6	5×10^4
(+) Methacholine	7.9×10^6	6×10^5	2×10^4
(–) Methacholine	4×10^4	6×10^3	1×10^3
r (+/–)	198	100	20
r/geomean affinity	3.5×10^{-4}	1.7×10^{-3}	4.5×10^{-3}

ceptor leads to a reduction in agonist binding affinity and changes in structure-activity relationships (unpublished). In Table 2 it can be seen that the affinity of both enantiomers of methacholine are reduced and the selectivity greatly reduced, but only in proportion to the loss of binding affinity, however after PCMB acetylcholine now binds less well than either form of methacholine.

Table 2. *Effect of PCMB on muscarinic receptors (IC_{50}^{-1})*

	Before	After treatment with PCMB
Acetylcholine	1.3×10^5	4.5×10^2
(+) Methacholine	9×10^4	1.5×10^3
(–) Methacholine	2.1×10^3	5.2×10^2
r (+/–)	43	2.9
r/geomean affinity	3.1×10^{-3}	3.3×10^{-3}

On the same kind of principle, the ratio between affinities of isomers can be used as a useful method of distinguishing receptor types when a drug is able to bind to more than one receptor. A good example of this is the case of neuroleptic drugs butaclamol and flupenthixol which exist as pairs of geometrical isomers. When their binding to cerebral cortex membrane fragments is studied in competition with various labelled ligands, as anticipated the ratios of activities are different[3] (Table 3). Of course this is again just a special case of structure-activity differences.

Table 3. *Binding constants on cerebral cortex receptors.*

	In competition with ^{3}H-labelled				
	Dopamine	5-HT	LSD	Haloperidol	WB4101
(+) Butaclamol	1×10^7	1×10^6	2×10^7	1×10^9	2.9×10^7
(–) Butaclamol	6.3×10^4	1.2×10^5	1.4×10^5	8×10^5	4×10^5
r (+/–)	160	8	143	1250	73
(α) Flupenthixol	5×10^6	2.5×10^5	1×10^7	5×10^8	1.4×10^8
(β) Flupenthixol	10^5	1.6×10^4	1.6×10^5	1×10^7	6.7×10^6
r (α/β)	50	16	63	50	21

There is little point in cataloguing the large number of instances in which differences in binding of stereoisomers has been reported, it is much more useful to seek information about the underlying molecular interactions. In some cases we can investigate these in a rather direct way by nuclear magnetic resonance. For instance, the enzyme dihydrofolate reductase (DHFR) will bind L-4-aminobenzoylglutamate (L-PABG) which is the terminal fragment of the normal substrate for the enzyme, dihydrofolate. In the complex we can identify two very clear effects, firstly the ortho and meta protons are shifted upfield by 0.41 and 0.58 ppm respectively, this is a ring current shift due to proximity of the benzoyl ring to an aromatic amino acid residue in the enzyme almost certainly Phe. Secondly, one of the histidines has its pK shifted and later evidence showed that this is due to interaction of the γ-carboxyl of the glumate with histidine 28 in the enzyme[4] (Table 4). By contrast, D-PABG binds only about one third as well as L-PABG and we find that there is no detectable shift of the benzoyl protons on binding. However, the effect on histidine 28 is the same as that of the L-PABG. The implication is that the dicarboxylic backbone of the PABG is binding in the same configuration for both isomers, but in D-PABG the direction of the bond at the α-position no longer permits the benzoyl group to take up the same position on the enzyme as in L-PABG. Indeed the lack of shift is compatible with the aminobenzoyl group remaining in the solvent.

Table 4. *Binding of the enantiomers of 4-amino benzoyl glutamic acid to DHFR*

	Binding constant M^{-1}	Δ (ppm) Ortho	Meta
L-PABG	1.05×10^3	-0.41 ± 0.02	-0.58 ± 0.02
D-PABG	0.34×10^3	< 0.05	< 0.05

The interpretation of this experiment seems to be that we can regard the binding of L-PABG as partitioned into energetic components attributable to the 4-aminobenzoyl and the dicarboxylic fragments. In D-PABG the changed stereochemistry has made it impossible for both fragments to bind simultaneously and we are then in an either/or situation in which the outcome is determined by the relative magnitudes of the binding energies of the two fragments. The experimental result strongly suggests that binding of the dicarboxylic acid is much the stronger and that the lower affinity of D-PABG is due to a loss of the binding of the 4-aminobenzoyl moiety. Considered in this simple way we can assess the binding of the dicarboxyly acid part as ~ 3.6 Kcals/mol and that of the 4-aminobenzoyl moiety as 0.9 Kcals/mol - thus in the 'competition' for binding the probability is about 80:1 in favour of the dicarboxylic acid part. It is obvious that this is a very crude argument that ignores, entropic and other effects, but it illustrates a useful principle.

Interesting information of a similar kind has come from the study of vancomycin, an antibiotic that exerts its action by binding the mucopeptides that are the substrates for the synthesis of the cell wall of many bacteria. These peptides terminate in a DAla-DAla sequence which is the essential feature which binds to the antibiotic, and is shown even in the simple blocked dipeptide, acetyl-DAla-DAla. The nmr shifts of the protons of the peptide can be readily monitored, and since the binding site of vancomycin is rich in aromatic groups, large ring current shifts occur which will depend on the orientation of the peptide. In Acetyl-DAla-DAla itself large upfield shifts are found for the methyl protons of both D-Ala residues and a small shift in the same direction for the acetyl methyl. Unfortunately Ac-DAla-Lala does not bind significantly, but Ac-Lala-Dala does bind although 50 times more weakly. The shift of the terminal Ala-CH_3 is quite unaltered, but neither the other Ala-CH_3

nor the acetyl-CH_3 show any detectable shift[5] (Table 5). Evidently both these groups are now distant from the aromatic groups in the vancomycin. In Ac-Gly-DAla, the D-Ala-CH_3 shows again the same shift and the Acetyl-CH_3 a larger shift than in Ac-DAla-DAla, and the binding constant is only modestly reduced. Here the acetyl group while not in exactly the same location as in the parent compound is evidently within the field of the vancomycin aromatics. A reasonable interpretation is that in Ac-LAla-DAla the steric inappropriateness of the L-Ala also leads to the inability of the acetyl group to enter the binding site. The pattern of reduced binding for the inappropriate analogue and a modest increase in binding for the appropriate analogue as related to Ac-Gly-Dala bears an interesting resemblance to the binding of leuenkephalin to the opiate receptor. Here when the 2-Gly residue is replaced by 2-D-Ala the binding is increased by a factor of 3, but replacement by 2-L-Ala reduces binding by 500 fold[6].

Table 5. *D-Ala peptides bound to vancomycin*

Binding constant (M^{-1})	Shifts in CH_3 protons (ppm)		
	Ac –	DAla –	DAla
14×10^3	– 0.09	– 0.20	– 0.57
	Ac –	LAla –	DAla
0.3×10^3	0	0	– 0.60
	Ac –	Gly –	DAla
6×10^3	– 0.18	–	– 0.55

Movement of a group into or out of the receptor surface as emphasised earlier is mainly a matter of energetics and this can be illustrated in a case where steric considerations are unlikely to be important. In the complex between the coenzyme NADP and DHFR we can monitor the shifts of the adenine and nicotinamide protons[7] and these can be seen in Table 6. There are large chemical shifts on all the protons and the differences are large enough to indicate that we should have excellent parameters for judging any changes in geometry. We will examine just two alterations in the molecule, the replacement of the carboxyamide group by carbothioamide, or by carbomethyl (acetyl). In the first of these thio NADP (TNADP), there is some change in the adenine A8 shift but the dramatic change in all the nicotinamide protons is the striking alteration. These shifts are all low and would be compatible with the interpretation that the nicotinamide group has swung out of the binding site into the solvent. There are also changes in the ^{31}P shifts of the pyrophosphate backbone and such shifts are attributable to changed bond angles and dihedrals around the phosphorus atoms and therefore confirm the change in the conformation of the bound coenzyme. The sulphur substitution decreases the binding energy by a factor of 4 due either to the reduced hydrogen binding capacity of the sulphur or to steric hindrance because of its greater bulk compared with oxygen.

In the second substitution acetylpyridine-NADP (APADP), the adenine proton shifts are unchanged but the nicotinamide shifts are not similar to those of TNADP but are selectively changed; there is only a small change at N_2 a reduction to 50% at N_4 and N_5 but complete loss of shift at N_6. It is evidently a more subtle alteration which probably involves some rotation about the nicotinamide axis. The phosphorus shifts also reveal a changed conformation in the pyrophosphate not identical to that in TNADP.

Table 6. *Shifts of ^1H and ^{31}P of coenzymes on binding to DHFR (ppm)*

	NADP	TNADP	APADP
	(pyridine ring, N–R, C(=O)NH₂)	(pyridine ring, N–R, C(=S)NH₂)	(pyridine ring, N–R, C(=O)CH₃)
A_2	-0.86	-0.82	-0.82
A_8	-0.39	-0.54	-0.54
N_2	0.61	-0.03	0.50
N_4	1.36	0.11	0.64
N_5	0.97	-	0.48
N_6	0.73	0.09	-0.03
PP-A	-0.16	+0.15	+0.35
PP-B	-1.73	-0.73	-0.83
Binding constant (M^{-1})	6.1×10^4	1.4×10^4	8.9×10^3

These changes are subtle and illustrate a general principle applicable to stereospecific binding that even small changes in ligand structure may lead to significant alterations in the binding mode.

An extreme and very interesting example of this principle is seen in methotrexate, which is an analogue of folic acid in which the oxygen atom at position 4 on the pteridine ring has been replaced by an amino group (there is also N-methylation at position 10 but this is not relevant to the argument). In simple chemical terms this substitution increases the basicity of the pteridine A ring so that at neutral pH the ring bears a positive change whereas in folate the ring is unchanged. The effect of this substitution is to increase the binding to DHFR by more than four orders of magnitude, but more interestingly to change a substrate into an inhibitor. This occurs without any major changes in the chemistry of the B ring which is the site of stereoselective hydrogen transfer in the action of the enzyme. A combination of crystallography and nmr[8,9] has shown that in the complex with the enzyme the pteridine ring of methotrexate is rotated 180° about the axis of the 6-group as compared with folate (Fig. 1). This allows the 4-amino group to form a useful association with leu 4 and Ala 97 in the enzyme. The orientation of the B ring is altered but it is still able to come into close enough relationship with the nicotinamide of NADPH so as to make hydrogen transfer to look possible. The failure to act as a substrate is therefore not explained.

A more complex stereochemical binding problem is seen in the binding of folinic acid to DHFR. This is a diastereomeric molecule with assymmetric centres at C-6 as well as at the α-position of the glutamate residue and in addition has a formyl group at N-5 which exists as cis-trans isomers. Commercial folinic acid is a mixture of the 6S-αS isomer which has a binding constant of $10^8 M^{-1}$ for DHFR and the 6R-αS isomer whose binding constant is only $10^4 M^{-1}$. We can measure directly in bound folinic acid the 3′5′ benzoyl protons which are shifted upfield to different extents for the 6R and 6S forms. In free folinic two signals are seen for the formyl group corresponding to the forms with the formyl oxygen trans or cis to the pteridine 4-oxo. These two isomers interconvert slowly on the nmr timescale[10] with the trans form being 3 times as abundant as the cis form. On binding 6S-αS folinic only the trans

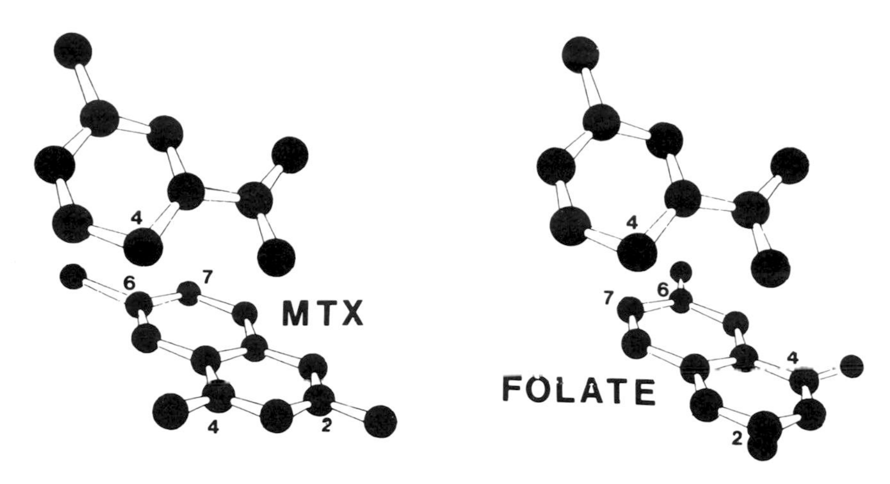

Figure 1. *Relationship of methotrexate and folate pteridine rings to the nicotinamide portion of NADP when bound to DAFR. Note the rotation of the pteridine ring by 180°.*

form is seen and the formyl proton is shifted downfield by 0.29 ppm. On the other hand bound 6R-αS folinic shows both cis and trans formyl groups in unchanged proportions from those in the free acid, and in this case the trans formyl group is shifted downfield by 0.10 ppm but the cis formyl group is shifted upfield by 0.13 ppm. In addition we can observe the histidines and tyrosines in the enzyme. Histidine-28 has its pK increased by 6S-αS folinic in the same way as by other folates and this is be due to an interaction with the γ-carboxyl of the glutamate. The effect of 6R-αS folinic seems to be identical. In addition both isomers have identical effects on the enzyme tyrosine resonances and these are known not to due to direct effects but to be exerted through conformational changes.

Our evidence thus suggests that the glutamate part of the stereoisomers is able to bind in essentially the same way in both isomers, the benzoyl ring is disturbed and the formyl group is bound in such a way that in the 6S-αS isomer binding of the cis formyl group is sterically prohibited, whereas for the 6R-αS isomer the formyl group cannot be directly interacting with the enzyme, since the equilibrium between cis and trans is not perturbed, but the groups are still sufficiently close to the enzyme to be significantly shifted on binding.

The evidence in this case then suggests that the PABG part of the molecule binds in a similar fashion in both isomers and that the major changes accounting for the 5.7 Kcal/mol difference in binding must be in the way the pteridine portion binds but this is only meagrely monitored by the formyl group.

In this discussion I have largely dealt with drug-receptor interactions as though the receptor was a fixed structure. However, the functioning of receptors as transducers depends on conformational transitions in the receptor. The differences between the subtypes of muscarinic receptors are believed to be related in some way to their conformational states and induced conformational changes in other receptors are accompanied by changes in structure-activity relationships. For this reason, we must always bear in mind that stereoisomers may be binding to difference conformational states of the receptor.

Summary

Stereoisomeric molecules cannot bind to the receptor with the same overall geometry. Binding of the unfavoured isomer may involve competition for binding between subregions of the molecule whose outcome depends on the respective energetic contributions to binding. Small alterations in ligand structure can lead to major alterations in binding geometry. Finally, it should be borne in mind that drug receptors are intrinsically conformationally labile. There may be cases where stereoisomers bind to different conformational states of the receptor.

References

1. S.H. Snyder, G.W. Pasternak and C.B. Pert in Handbook of Psychopharmacology, Ed. L.L. Iversen, S.D. Iversen and S.H. Snyder. Plenum Press 5, 329 (1975).
2. N.J.M. Birdsall, E.C. Hulme and A.S.V. Burgen. Proc. R. Soc. Lond B. 207 1 (1980).
3. S.J. Enna, J.P. Bennett, D.R. Burk, I. Creese and S.H. Snyder, Nature, 263 338 (1976).
4. G.C.K. Roberts, J. Feeney, A.S.V. Burgen, V. Yuferev, J.G. Dann and R. Bjur. Biochemistry, 13 5351 (1974).
5. J.P. Brown, L. Terenius, J. Feeney and A.S.V. Burgen. Mol. Pharmacol., 11 126 (1975).
6. C.R. Beddell, R.B. Clark, G.W. Hardy, L.A. Lowe, F.B. Ubatuba, J.R. Vane and S. Wilkinson. Proc. R. Soc. Lond. B 198 249 (1977).
7. E.I. Hyde, B. Birdsall, G.C.K. Roberts, J. Feeney and A.S.V. Burgen. Biochemistry 19 3738 (1980).
8. D.A. Matthews, R.A. Aldin, J.T. Bolin, D.J. Filman, S.T. Freer, R. Hamlin, W.G.J. Hol, R.L. Kisliuk, E.J. Pastore, L.T. Plante, N. Xuong and J. Kraut. J. Biol. Chem. 254 6946 (1978).
9. P.A. Charlton, D.W. Young, B. Birdsall, J. Feeney and G.C.K. Roberts. J. Chem. Soc. Chem. Comm. 922 (1979).
10. J. Feeney, B. Birdsall, J.P. Albrand, G.C.K. Roberts, A.S.V. Burgen, P.A. Charlton and D.W. Young. Biochemistry 20 1837 (1981).

Advantages and Disadvantages in the Application of Bioactive Racemates or Specific Isomers as Drugs

W. Soudijn

Abstract

Many drugs having a centre of asymmetry are used in clinical practice as racemates, although it seems good sense to use the more active isomer (eutomer) instead, in order to decrease the load of xenobiotics on the organism and to escape from the risk of unwanted, toxic side effects caused by the presence of the less active or inactive isomer (distomer) or its metabolites. There are however instances where it is better to use racemates rather than eutomers; e.g. when racemates are more active, less toxic, of longer (shorter) duration of action than the eutomers. More attention should be given to the pharmacological, metabolic, kinetic and toxicological properties of both isomers separately and in combination as racemates.
The obvious question whether it would not be wiser to design effective drugs without centres of asymmetry will be dealt with briefly.

Introduction

Medicinal chemistry has generated many drugs with one or more centres of asymmetry or showing atropisomerism or cis-transisomerism. For this paper I would like to borrow the terms eutomer for the isomer with the highest activity (Greek eu = good) and distomer for the isomer with the lowest activity (Greek dis = bad) for the same pharmacological action, from Ariëns, Lehmann and de Miranda who coined them in about 1976.

Many drugs used in clinical practice are racemates because often the animal and clinical pharmacology, the toxicology and since the thalidomide disaster, the teratology was studied with the racemates as resolution into eutomer and distomer proved to be difficult, costly or at the time of discovery of the drug even impossible with the techniques then available. Sometimes drugs of the same pharmacological class are marketed as racemates for apparently no compelling reasons other than economic ones.

For instance from the twelve β-adrenergic blocking agents – used for the treatment of hypertension and angina pectoris – only two, penbutolol and timolol are marketed as the eutomers, the majority as racemates while the β-blocking activity resides mainly in the levo (–) isomers.

The antihypertensive agent methyldopa (α-methyl-3,4-dihydroxyphenyl-alanine) is a prodrug which is decarboxylated and β-hydroxylated to the active compound α-methylnorepinephrine (Fig. 1). Only the l-isomer produces the antihypertensive effect. The d-isomer may only contribute to the toxicity of the racemate. The drug is marketed by different companies as the levo isomer or as the racemate.

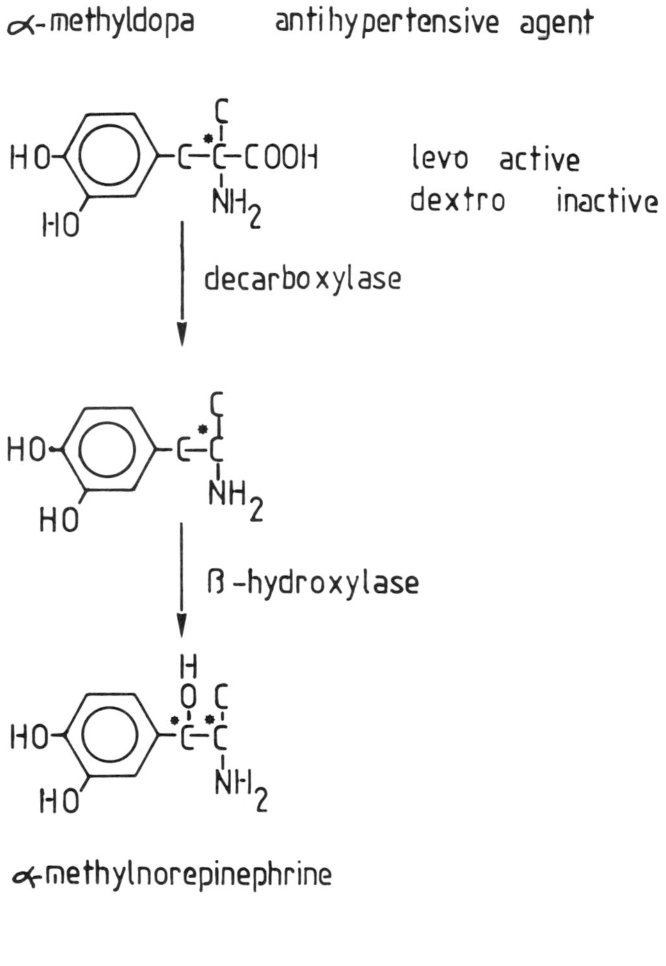

α-methyldopa antihypertensive agent

levo active
dextro inactive

decarboxylase

β-hydroxylase

α-methylnorepinephrine

Figure 1.

Since we are living in times of Good Practice like Good Laboratory Practice and Good Manufacturing Practice this is perhaps the right moment to ask – tentativily of course – whether the clinical use of racemates instead of eutomers is really Good Pharmacotherapeutical Practice, or should we use the eutomers exclusively?

It is obvious that eutomers should be used when:

- *The distomer contributes to the provoking of side effects.*

  (±) Ketamine HCl (Fig. 2) is a parenteral anesthetic that lacks respiratory depression but the rather high incidence of side effects like post operative restlessness, combativeness, loss of selfcontrol, agitation and disorientation have limited its clinical usefulness. White et al.,[1] tested the effects of the racemate as well as both isomers in surgical patients in a randomized double blind fashion and found that the (+) isomer HCl salt gave a more adequate anesthesia (less adjunctive agents necessary) and considerably less disturbing side effects than both the racemate and the (–) isomer HCl salt (note that the (–) HCl salt = (+) base) at an induction dose of 1 mg/kg versus 2 mg/kg for the racemate and 3 mg/kg for the distomer.

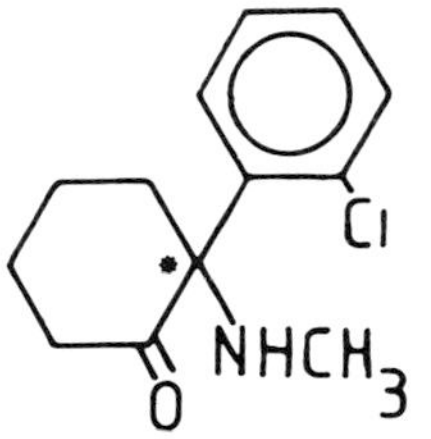

·HCl = KETAMINE parenteral anesthetic (±)

NHCH$_3$

Cl

O

patients:

hypnotic, analgesic: $(+) > (\pm) > (-)$

unwanted sideeffects: $(+) < (\pm) < (-)$

HCl salt

(+) salt = (–) base

cross contamination: 10 – 20 %

P. F. WHITE et al. Anesthesiology 52 231 (80)

Figure 2.

– *The distomer counteracts the pharmacological action of the eutomer.*
Etozoline, a cis compound, is a loopdiuretic that is metabolized to the active ozolinone, an experimental drug (Fig. 3). The (–) isomer is the eutomer with the diuretic properties, the (+) isomer however is not a diuretic but inhibits the diuretic action of the eutomer and of furosemide in the rat[2].

In barbiturates used as anesthetics and hypnotics the neuronal depressant activity generally predominates in the S(–) isomer while in the R(+) isomer the excitatory effect predominates. An extreme example is 5-ethyl, 5-(1,3-dimethylbutyl)barbiturate (Fig. 4) where the S(–) isomer is a depressant, the R(+) isomer a convulsant and the racemate a convulsant drug unsuitable for anesthetic purposes[3]. In mice the S(–) isomers of pentobarbital, secobarbital, thiopental and thioamylal are more toxic than the R(+) isomers or the racemates. The therapeutic safety in mice – i.e. the ratio LD_{50}/ED_{50} – however appears to be the same, about 3,4 for S(–), R(+) or RS of all compounds[4].
Several anesthesiologists seem to prefer the use of racemates because they consider the presence of a convulsant component in the racemate as a safeguard in cases of accidental overdosing. I do not know whether this consideration is based on sound experimental data, experience or faith, but I must confess that I

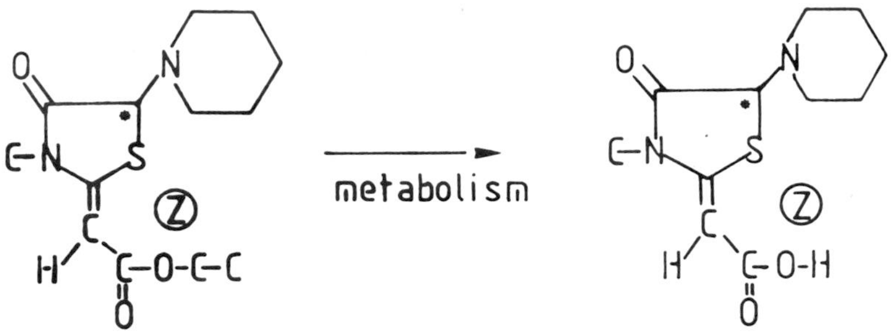

metabolism

etozoline —— loop diuretics —— ozolinone

(−) = diuretic

(+) = not diuretic, inhibits low dose (−) and furosemide

J. GREVEN et al, Pflügers Arch. 384 57 (80)

Figure 3.

barbiturates hypnotics anesthetics

in general: S (−) depressant effect predominates

R (+) excitatory effect predominates

extreme:

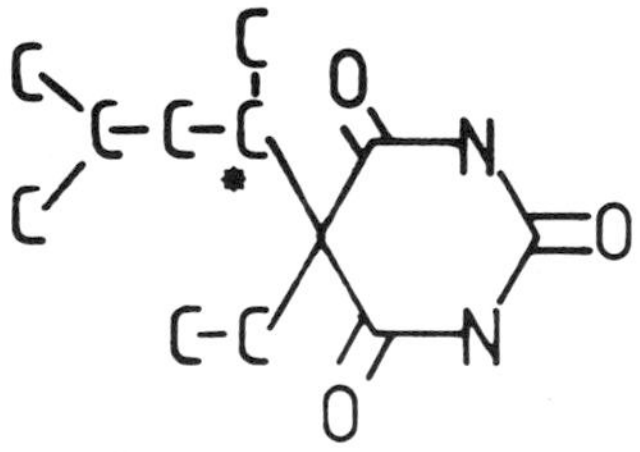

S (−) depressant

R (+) convulsant

RS convulsant

I.K. HO Ann. Rev. Pharmacol. Tox. 21 83 (81)

Figure 4.

have my doubts.

The narcotic analgesics seem to be an exception to the rule that eutomers should be used when the distomer counteracts the pharmacological action of the eutomer.

Recently a new narcotic analgesic, picenadol (LY 150720) - presently undergoing clinical trials - was reported by Zimmerman[5] (Fig. 5). The racemate is a partial agonist because the d-isomer is a potent morphinomimetic while the l-isomer is a narcotic antagonist.

picenadol narcotic agonist - antagonist

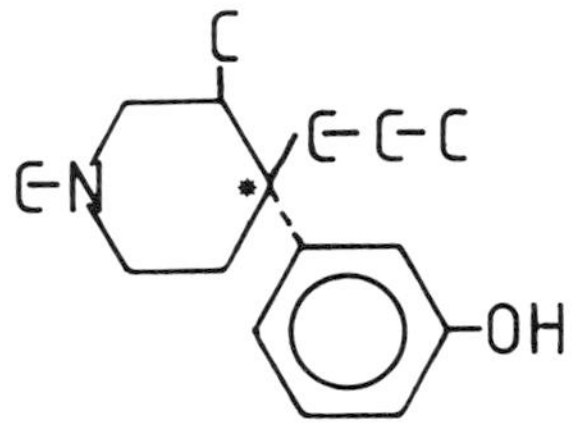

d = potent agonist

l = antagonist

d l = partial agonist

D.M.ZIMMERMAN et al. Ann.Rpts.Med.Chem. 17 21 (82)

Figure 5.

- *The distomer is metabolized to a product with an unfavourable pharmacological action.*
  Deprenyl, a monoaminooxydase inhibitor is used in the treatment of depression (Fig. 6). The (-) isomer is a much more potent mao-b inhibitor than the (+) isomer.
  The metabolic products of the racemate are dextro and levo metamphetamine[6]. As the dextro amphetamines are pharmacologically more potent than the levo amphetamines, it was to be expected that unfavourable symptoms caused by the dextro amphetamines should occur in patients treated with racemic deprenyl. This proved indeed to be the case and racemic deprenyl in clinical practice has since been replaced by (-) deprenyl.
- *The distomer is metabolized to toxic products.*
  The local anesthetic potencies of the R(-) and S(+) isomers of prilocaïne (Fig. 7) do not differ very much so in this sense it is difficult to conclude which one is the eutomer. However racemic prilocaïne may cause methemoglobinemia and the clinical use is therefore declining. Prilocaïne, and especially the R(-) isomer, is metabolized to an aniline (orthotoluidine) and to the corresponding para- and ortho aminophenols, products that are higly toxic and responsible for the methemoglobinemia[7]. In this sense we could call the R(-) isomer a distomer. If in clinical practice the S(+) isomer had been used the incidence of methemoglobinemia would have been nil or very low indeed.
  Another example, which is rather speculative I fear, but useful to illustrate the point is thalidomide (Fig. 8). It was shown in recent years by Blaschke[8] and Ockenfels[9,10] that S(-) thalidomide but not R(+) thalidomide is transformed in vivo into l-N-phtaloylglutamine and l-N-phtaloylglutamine acid, products that are both embryotoxic and teratogenic in SWS mice and in Natal rats. R(+) thalidomide is transformed into d-N-phtaloylglutamine and d-N-phtaloylglatamic acids which are neither embryotoxic nor teratogenic in these species. If lengthening of hexobarbital induced sleep in mice is an appropriate test for hyp-

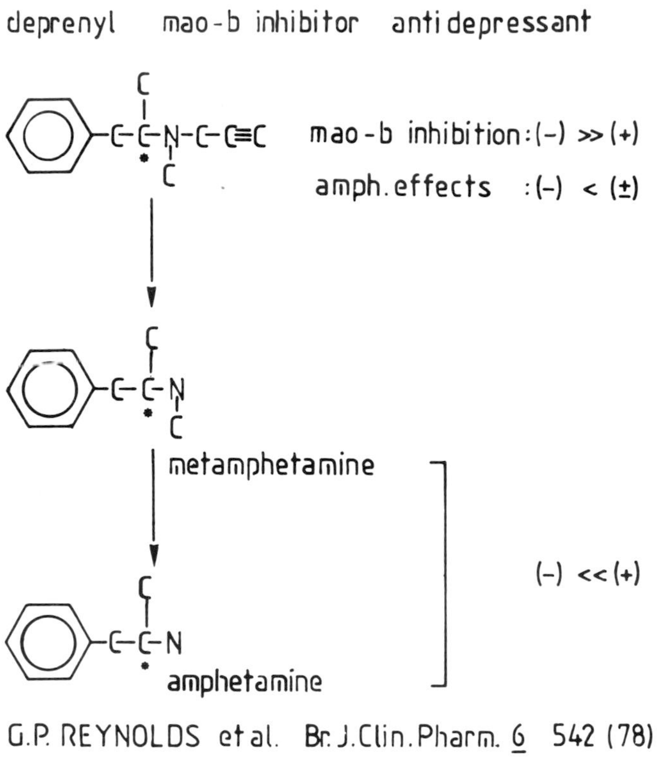

deprenyl mao-b inhibitor antidepressant

mao-b inhibition: (-) >> (+)

amph. effects : (-) < (±)

metamphetamine

amphetamine

(-) << (+)

G.P. REYNOLDS et al. Br. J. Clin. Pharm. 6 542 (78)

Figure 6.

notic potency, then according to Fabro[11] the racemate of thalidomide, its R(+) and S(-) isomers are equiactive, and if in addition the neurotoxicity of racemic thalidomide that may occur when high doses are used over long periods, resides in the S(-) isomer and not in the R(+) isomer, a supposition for which no published data as far as I know, is available then R(+) thalidomide might be a useful hypnotic drug. These are – as said earlier – mere speculations of course for if e.g. the metabolites of R(+) thalidomide are enzymatically racemized or inverted the game is off.

So far it is obvious that in clinical practice the use of eutomers is to be preferred to the use of racemates. The choice becomes a little more difficult when the distomer and its metabolites seem, from a pharmacological and toxicological view, fairly harmless and do not interact in any way with the action and activity of the eutomer. Now the opinions will tend to differ. Some people will say: when the distomer does not cause any harm, why go to the trouble of resolution of the racemate as it will only raise the price.

Other more prudent or timid people will say: The benefit of the presence of the distomer is nil but the risk is unpredictable because:

a) The distomer or its metabolites may interact with a drug of comedication thereby causing trouble.
b) There is no way of knowing what the distomer or its metabolites will do when the patients do have other diseases together with the one for which they are treated with the racemate.
c) some patients-admittedly a small percentage – could be allergic to the distomer and not to the eutomer.

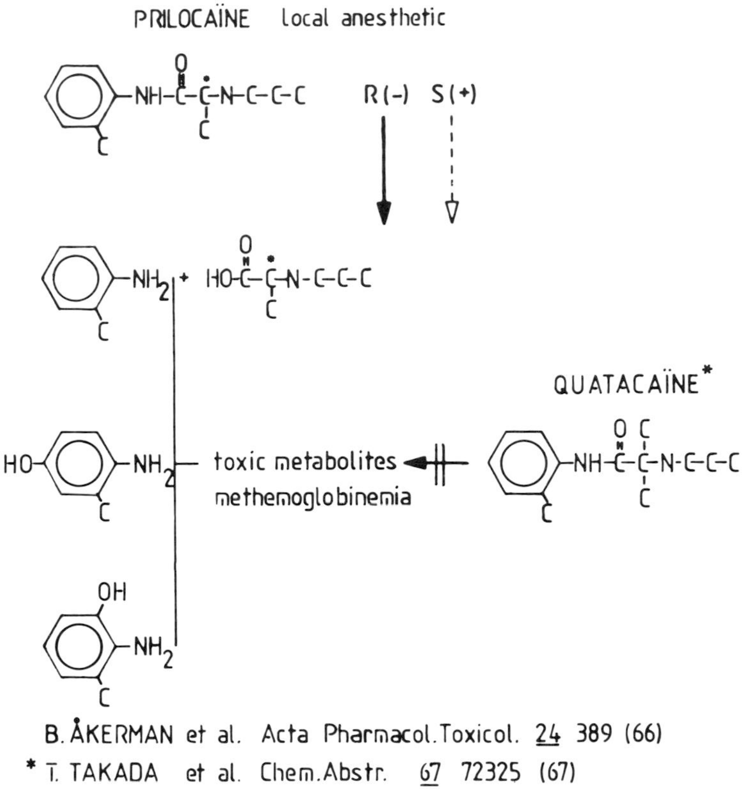

PRILOCAÏNE Local anesthetic

R(-) S(+)

toxic metabolites methemoglobinemia

QUATACAÏNE*

B. ÅKERMAN et al. Acta Pharmacol. Toxicol. 24 389 (66)

* T. TAKADA et al. Chem. Abstr. 67 72325 (67)

Figure 7.

So why take chances you do not have to take at all when using the eutomer instead of the racemate. And the cynics will side with the prudent this time and think: Well OK, higher prices, higher profits. So it depends on your psychological make-up whose side you are on but I think everybody will agree that it is a laudable principle to try and keep the load of xenobiotics as low as possible. However this does not imply that racemates should never be used. Racemates should be used when:

- *Racemate and eutomer are equipotent and equitoxic*

  In Fig. 9 two experimental narcotic analgesics are shown, one with the usual potency ratio of racemate over eutomer of about two, the other - a methylsubstituted analog - with the rather unusual potency ratio of one in the rat tail radiant heat test after subcutaneous administration[12]. This could be explained by assuming that the weakly active distomer inhibits the biotransformation and thus inactivation of the eutomer. If the same holds true in man and if racemate and eutomer are equitoxic there is hardly any reason for not using the racemate. The antimaterial chloroquine diphosphate (Fig. 10) offers a more complicated example. Haberkorn[13] showed that oral administration of 7.5 mg d-isomer during four days to mice infected with Plasmodium bergheï results in a 50% cure rate. Racemate and l-isomer proved to be inactive at that dose. A 50% cure rate however is not acceptable in clinical practice. At 70-100% cure rates the difference in potency between eutomer and racemate has disappeared. Although in acute experiments distomer and racemate are more toxic than the eutomer, the

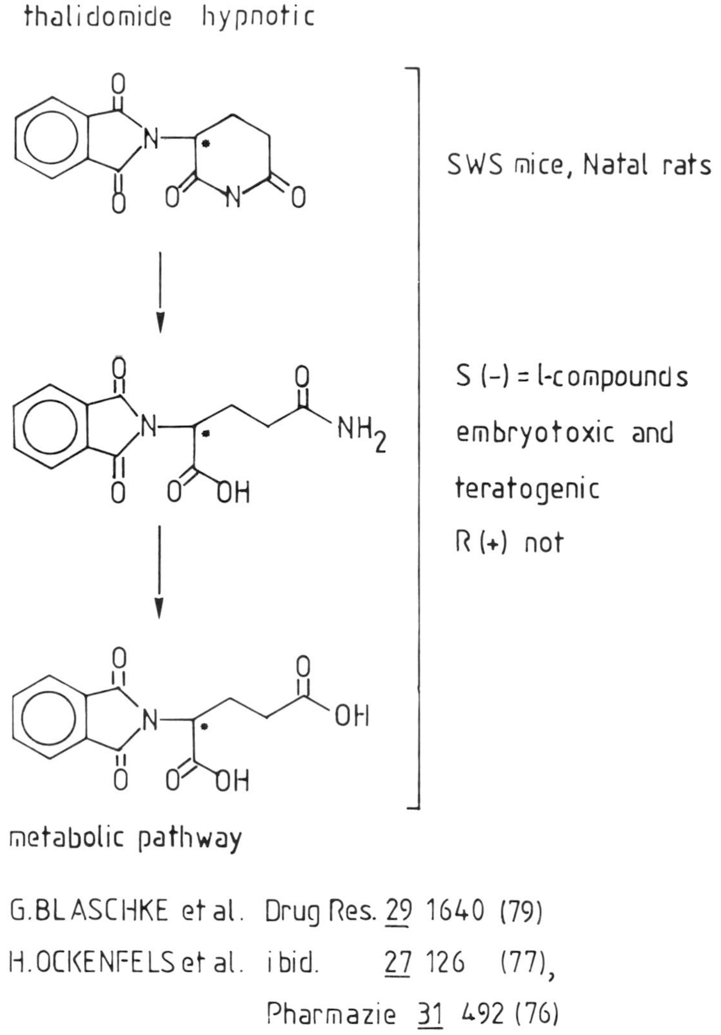

thalidomide hypnotic

SWS mice, Natal rats

S (-) = l-compounds embryotoxic and teratogenic

R (+) not

metabolic pathway

G.BLASCHKE et al. Drug Res. 29 1640 (79)

H.OCKENFELS et al. ibid. 27 126 (77),

Pharmazie 31 492 (76)

Figure 8.

difference in toxicity grows less in 'chronic' (four days) experiments. The toxicity differences may virtually disappear when the treatment period is extended. As malaria in man requires long term treatment it is quite possible that the use of the eutomer instead of the racemate offers no advantages at all, if we may extrapolate the data from the animal experiments to man that is.

- *Racemate is more active than both intrinsically equipotent isomers.*
  The antihistaminic activity of racemate and isomers of isothipendyl (Fig. 11) is the same when tested in vitro in the guinea-pig ileum. However when tested orally for histamine protection in the guinea-pig it appears that the racemate is 1.4 times more potent than the d-isomer and 2.5 times more potent than the l-isomer[14]. Explanations of this phenomenon are probably to be found in the domain of pharmacokinetics. Differences in absorption, bioavailability, inhibition of biotransformation of the d-isomer by the l-isomer are all possible causes.

narcotic analgesics
rat tail radiant heat

	ED_{50} mg / kg sc	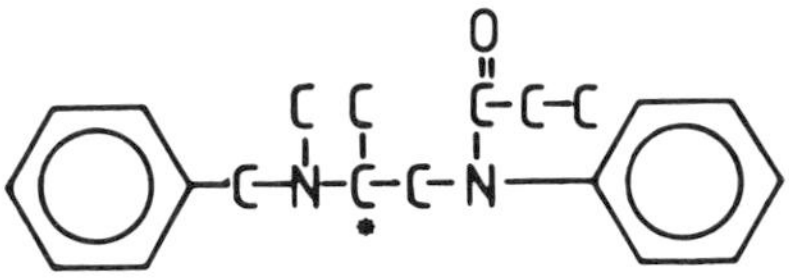
RS	8.0	
S (+)	4.3	RS / S ~ 2
R (-)	> 50	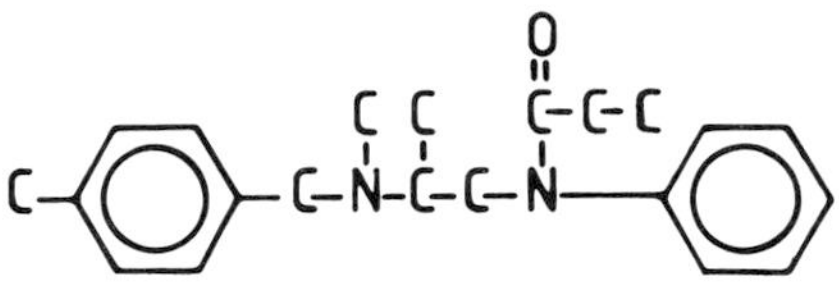
RS	1.6	Δ n.s.
S (+)	1.4	RS / S = 1
R (-)	> 50	

Ph. S. PORTOGHESE Th. N. RILEY J. Pharm. Sci. 54 1831 (65)

Figure 9.

CHLOROQUINE DIPHOSPHATE antimalarial

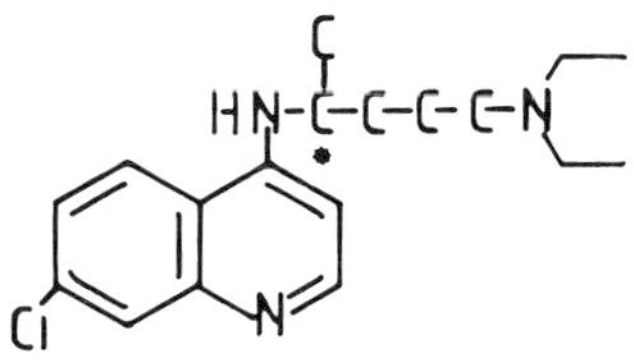

or dd 4 days % cure rate mice

mg / kg	d	dl	l	Pl. bergheï
7.5	50	0	0	
10	55	47	5	
20	70	70	0	
25	100	100	80	

toxicity (LD_{50}) ratio

	d/dl	dl/l	d/l
acute	1.4	1.3	1.8
"chronic" = 4 days	1.2	1.1	1.3

A. HABERKORN et al. Tropenmed. Parasit. 30 308 (79)

Figure 10.

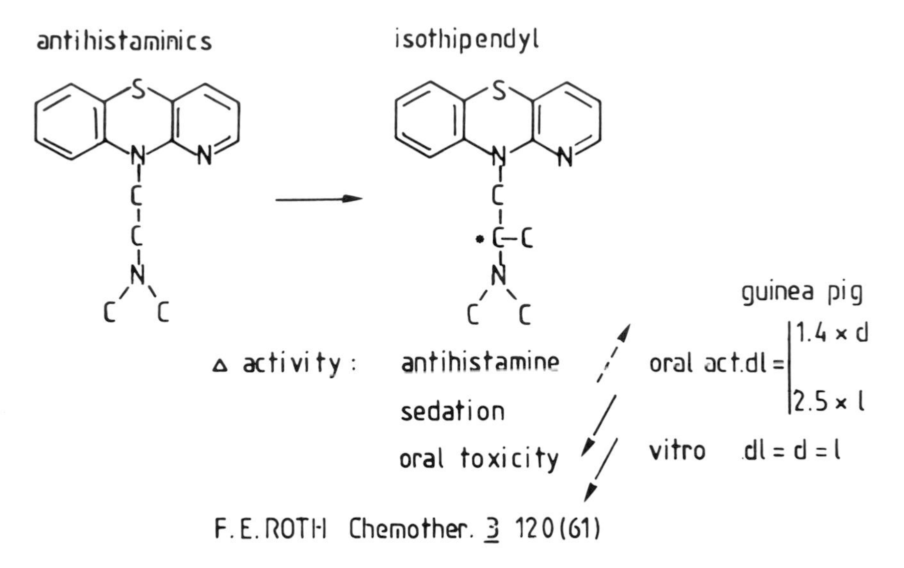

Figure 11.

Enzymatic inversion of the distomer.
Many anti-inflammatory drugs are phenylpropionic acid derivatives. In vitro the S(+) isomers are active inhibitors of the prostaglandin synthesis while the R(–) inhibitors are inactive. It was found that in several animal species including man the S(+) and the R(–) isomer are equipotent anti-inflammatory drugs. The explanation was found when it was demonstrated that the R(–) isomer (distomer) is inverted in vivo into the S(+) isomer (eutomer). An extreme example is clidanac[15] (Fig. 12). After oral administration to guinea-pigs the distomer R(–) was inverted for 90% into the eutomer S(+) within 3 hours. When the racemate was given 90% eutomer was formed in one hour.
The eutomer when administered was inverted for only about 5% into the distomer. There was little biotransformation and the drugs were excreted largely unchanged.

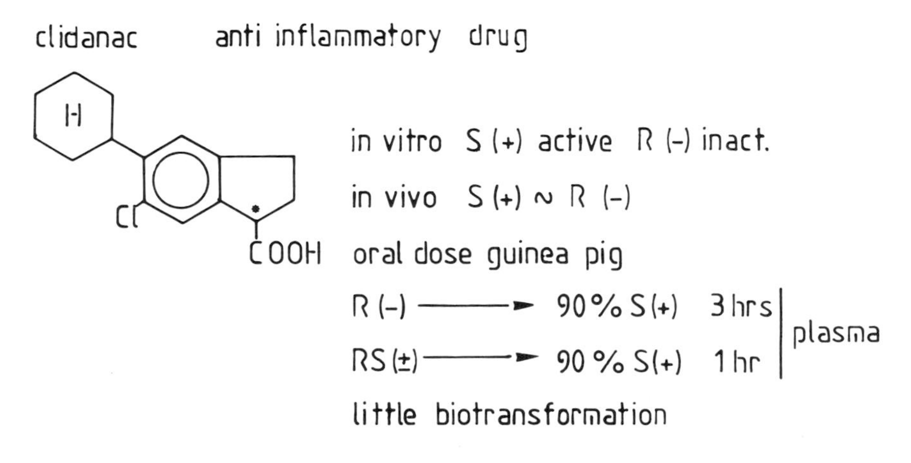

Figure 12.

- *The distomer counters the side effects of the eutomer.*
  Many of the diuretics currently used in pharmacotherapy lead to raising of the blood uric acid level through retention of urate. Hyperuricemia which may result from urate retention is a risk factor in cardiovascular disease and may induce nephropathy. Fanelli[16] showed that the racemate of the indacrynic acid analogue depicted in Fig. 13 is a uricosuric diuretic in chimpanzees and that the diuretic action is caused by the (+)-isomer while the uricosuric action resides in the (-)-isomer.

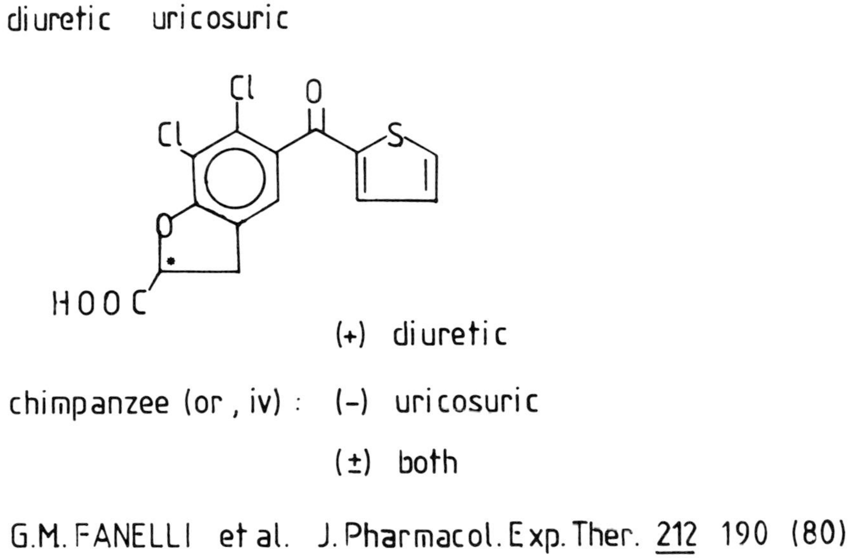

diuretic uricosuric

Cl O Cl S O HOOC

(+) diuretic

chimpanzee (or, iv) : (-) uricosuric

(±) both

G.M. FANELLI et al. J. Pharmacol. Exp. Ther. 212 190 (80)

Figure 13.

In contrast to other β-adrenergic agonists dobutamine (Fig. 14) enhances the contractile force of the heart at doses that do not provoke an increase in heart rate. This is the result of the combination of pharmacological actions of the (+)- and the (-) isomer. The β-agonistic properties of the racemate are mainly caused by the (+)-isomer while the α-adrenergic agonistic component in its pharmacological profile is caused by the (-)isomer[17].

The racemate is a useful drug and the cost of eutomer production is prohibitive.
It is sometimes difficult to decide from pharmacological data whether it is more sensible to use the racemate or one of the isomers, when the pharmacogical profile of the drugs is complex. The pharmacological profile of the isomers and racemate of the antidepressant drug oxaprotiline (Fig. 15) was studied in the rat by Waldmeier et al.[18] and Mishra et al.[19]. The S(+) isomer proved to be equiactive with the racemate in blocking the norepinephrine uptake and the down regulation of β-adrenergic receptors in the brain while the R(-) isomer proved to be weakly active in vitro and inactive in vivo experiments.
The α_1-adrenoceptor blocking potency of the R(-) isomer is somewhat higher than that of the S(+) isomer. The (+) and (-) isomers are virtually equipotent in tests for antihistamine and antiagressive properties.
The racemate is inactive in inhibiting the uptake of serotonin. It seems that only clinical pharmacology can provide the answer to the question which one of the three drugs, - racemate, S(+) isomer or maybe even the R(-) isomer - is to be

dobutamine β-agonist pos. inotropic

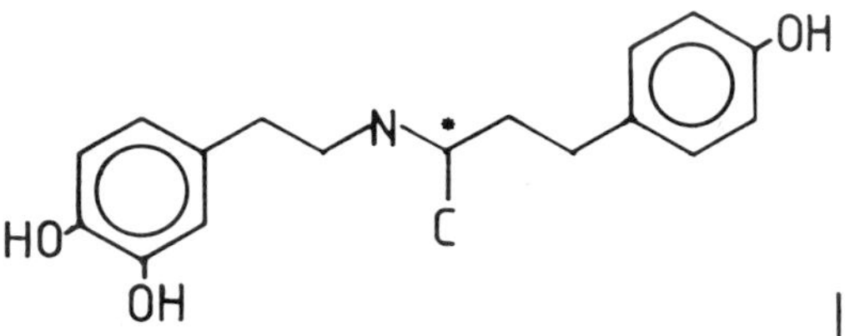

HO, OH, OH, N, C

adrenergic act. | (−) β-agonist + α-agonist
| (+) β-agonist + α-antagonist

Σ (±) β-agonist + α-agonist

R.R. RUFFOLO et al. J. Pharmacol. Exp. Ther. 219 447 (81)

Figure 14.

oxaprotiline anti depressant

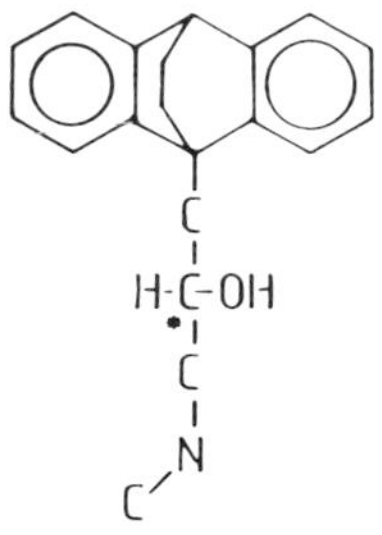

C, H-C-OH, C, N, C

rat brain | in vitro
| ex vivo | acute
| chronic

β-down regulation | S(+) ∼ (±) : R(−) weak - inactive
ne uptake block |

5-ht uptake block (±) inact. α_1-block (+) < (−)

anti histamine | (+) ∼ (−) clinically (±) ? (+) ? (−)
anti agressive |

P.C. WALDMEIER et al. Biochem. Pharmacol. 31 2169 (82)
R. MISHRA et al. Life Sci. 30 1747 (82)

Figure 15.

preferred in the treatment of depressive illness.

- *'Symmetry' or asymmetry?*

  From all that has been said thus far it seems evident that racemate and both enantiomers are usually three different pharmacological entities and that it requires extensive pharmacological, toxicological and clinical pharmacological research before we can decide whether it is advantageous to use racemates or enantiomers in clinical practice. These research efforts could be reduced to about one third when drugs without centres or planes of asymmetry could be developed with the same or higher efficacy than the asymmetric ones.

For example: the majority of narcotic analgesics have one or more centres of asymmetry. The potency difference between eutomer and distomer is usually large. Fentanyl and sufentanil belong to the most potent analgesics known and are without centre of asymmetry (Fig. 16).

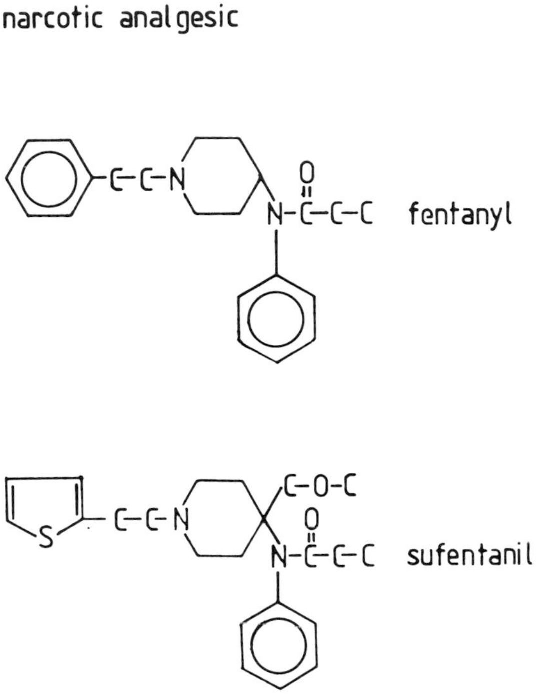

narcotic analgesic

fentanyl

sufentanil

Figure 16.

Takada et al.[20] claim that when the centre of asymmetry in prilocaïne is abolished by substitution with another methyl group (Fig. 7, quatacaïne) the local anesthetic potency and the duration of action is raised and toxic metabolites causing methemoglobinemia are not formed.
Most neuroleptic drugs are without centres of asymmetry and the neuroleptics with centres of asymmetry do not offer marked advantages in the treatment of schizophrenia, however the enantiomers are of course important research tools in the study of mechanisms of action and structure-activity relationship. On the other hand the introduction of a centre of asymmetry can be useful to improve a drug. For example: isothipendyl (Fig. 11) is an improvement of the drug without its centre of asymmetry because – although the increase of antihistaminic potency is moderate – oral toxicity and sedation decrease markedly (see for structure-activity relationship von Schlichtegroll[21]). Toldy[22] showed that the introduction of an asymmetric centre as in promethazine (Fig. 17) results in an increase of antihistaminic potency of over 100 percent. The asymmetric centre is not essential for antihistaminic action or for toxicity as both enantiomers are equipotent and equitoxic.

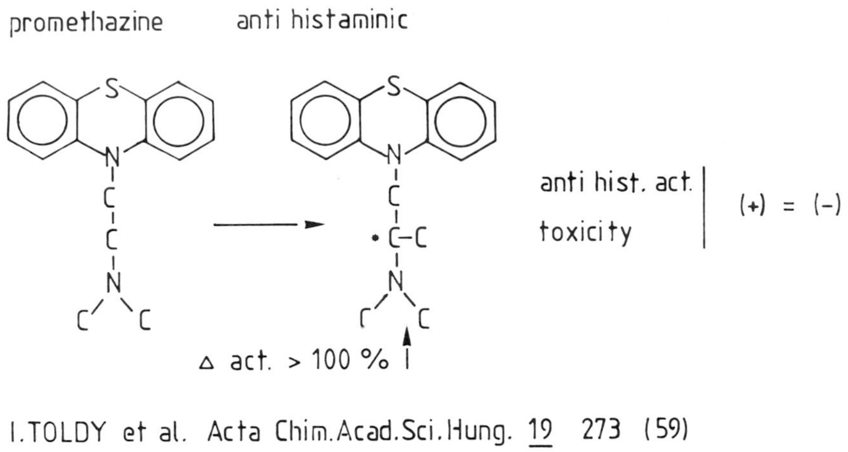

promethazine anti histaminic

anti hist. act. | toxicity | (+) = (−)

Δ act. > 100 %

I.TOLDY et al. Acta Chim.Acad.Sci.Hung. 19 273 (59)

Figure 17.

In conclusion one could try and draw a list of preference for the clinical use of drugs:

a) Drugs without a centre of asymmetry
b) Eutomers instead of racemates
c) Racemates when they have been proven superior to a or b,

while always bearing in mind the quip of H.L. Mencken:
'For every complex there is a solution that is neat, simple and wrong'.

References

1. P.F. White, J. Ham, W.L. Way, A.J. Trevor, Anesthesiology, *52*, 231 (1980).
2. J. Greven, W. Defrain, G. Glaser, K. Meywald, O. Heidenreich, Pflüger's Arch., *384*, 57 (1980).
3. I.K. Ho, R. Adron Harris, Ann. Rev. Pharmacol. Toxicol., *21*, 83 (1981).
4. H.D. Christensen, I. Sohee, Toxicol. Appl. Pharmacol, *26*, 495 (1973).
5. D.M. Zimmerman, P.D. Gesellchen, Ann. Rpts. Med. Chem., *21* (1982).
6. G.P. Reynolds, J.D. Elsworth, K. Blau, M. Sandler, A.J. Lees, G.M. Stern Br. J. Clin. Pharmac., *6*, 543 (1978).
7. B. Åkerman, S. Ross Acta Pharmacol. Toxicol., *24*, 389 (1966).
8. G. Blaschke, H.P. Kraft, K. Fickentscher, F. Köhler, Drug Res., *29*, 1640 (1979).
9. H. Ockenfels, F. Köhler, W. Meise, Drug Res., *27*, 126 (1977).
10. H. Ockenfels, F. Köhler, W. Meise, Pharmazie, *31*, 492 (1976).
11. S. Fabro, R.L. Smith, R.T. Williams, Nature, *215*, 296 (1967).
12. Ph.S. Porthogese, Th.N. Riley, J. Pharm. Sci., *54*, 1831 (1965).
13. A. Haberkorn, H.P. Kraft, G. Blaschke, Tropenmed. Parasit., *30*, 308 (1979).
14. F.E. Roth, Chemotherap., *3*, 120 (1961).
15. S. Tamura, S. Kuzuma, K. Kawai, S. Kishomoto, J. Pharm. Pharmacol., *33*, 701 (1981).
16. G.M. Fanelli, L.S. Watson, D.L. Bohn, H.F. Russo, J. Pharmacol. Exp. Therap., *212*, 190 (1980).
17. R.R. Ruffolo, T.A. Spradlin, G.D. Pollock, J.E. Waddell, P.J. Murphy, J. Pharmacol. Exp. Therap., *219*, 447 (1981).
18. P.C. Waldmeier, P.A. Baumann, K. Hauser, L. Maitre, A. Storni, Biochem. Pharmacol., *31*, 2169 (1982).
19. R. Mishra, D.D. Gillespie, R.A. Lovell, R.D. Robson, F. Sulser, Life Sciences, *30*, 1747 (1982).
20. T. Takada, M. Tada, A. Kiyomoto, Chem. Abstr., *67*, 72325 (1967).
21. A. von Schlichtegroll, Arzneimitt. Forsch., *7*, 237, *8*, 489 (1957) (1958).
22. I. Toldy, L. Vargha, I. Toth, J. Borsy, Acta Chim. Acad. Sci. Hung., *19*, 273 (1959).

Stereoselectivity in Adrenergic Agonists and Adrenergic Blocking Agents

R. R. Ruffolo, Jr.

Introduction

Stereoselectivity for adrenergic agonists and antagonists is an extremely complex subject since the adrenergic neuroeffector junction, unlike other neuroeffector junctions, contains many stereoselective processes. A hypothetical adrenergic neuroeffector junction is presented in Fig. 1 depicting several of the many processes known to occur at this site.

Postsynaptic α_1, β_1 and β_2 adrenergic receptors have been known for many years. Recently, postsynaptic α_2-adrenergic receptors have also been identified[1]. These four adrenergic receptor subtypes (process 1) possess their own particular stereo-

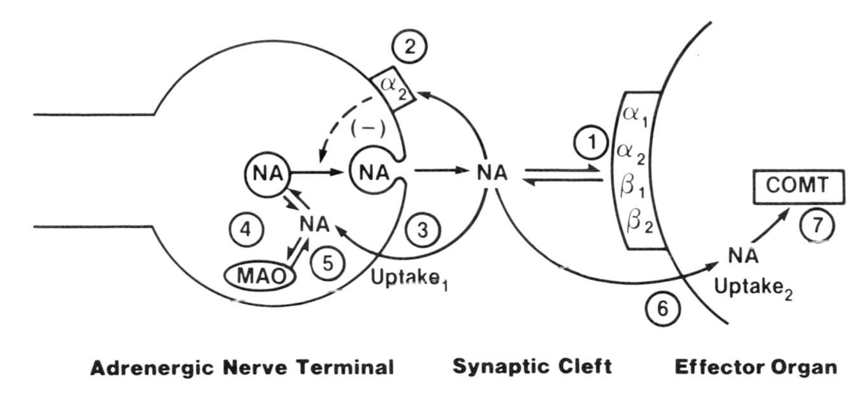

Figure 1. *Hypothetical adrenergic neuroeffector junction depicting several processes known to occur at this site. Many of these processes are stereoselective.*

chemical requirements for the natural neurotransmitter as well as for exogenously administered drugs. The existence of a presynaptic α_2-adrenergic receptor which regulates neurotransmitter liberation *via* a negative feedback system (process 2) is now known[2], and this has also been the target of recent stereochemical investigations[3-8]. Most of the natural neurotransmitter, noradrenaline, which is liberated by adrenergic nerve terminals, is removed from the synaptic cleft by the amine uptake pump ($uptake_1$)[9], which many believe to be a stereoselective process (process 3). In addition, cytoplasmic noradrenaline in the sympathetic nerve terminal is rapidly accumulated by adrenergic storage vesicles (process 4) which display high degree of stereoselectivity. Enzymatic inactivation of noradrenaline by monoamine oxidase (MAO), located predominantly in the cytoplasm of the sympathetic nerve terminal (process 5), and by catechol-O-methyltransferase (COMT), an extraneuronal enzyme (process 7), have been investigated for their stereochemical requirements as has extraneuronal uptake ($uptake_2$) (process 6). One must consider the configurational requirements of each of these varied processes for adrenergic drugs possessing either one or two asymmetric centers. In addition, conformational requirements for these different processes must also be addressed. For directly-acting drugs, the stereochemical requirements at the level of the postsynaptic and presynaptic adrenergic receptors are most critical, although $uptake_1$, $uptake_2$, MAO and COMT may all result in loss of activity. For indirectly-acting sympathomimetic amines, which work through liberation of endogenous stores of catecholamines, stereoselective considerations of $uptake_1$ and uptake into adrenergic storage vesicles are most critical since these sites regulate the access of the indirectly-acting sympathomimetic amine to those compartments within the sympathetic nerve terminal from which the neurotransmitter is liberated.

Stereochemical requirements for directly-acting adrenergic agonists

1. Configurational requirements for directly-acting adrenergic agonists.

a. Phenethylamines with asymmetry at the β-carbon atom: the Easson-Stedman Hypothesis. The most important theory governing the adrenergic activity of phenethylamines possessing one point of asymmetry at the β-carbon atom is the Easson-Stedman Hypothesis[10-12] (for reviews see [13-18]). This hypothesis proposes that a three-point attachment is involved in the binding of a sympathomimetic amine possessing an asymmetric β-carbon atom to what was then simply called the adrenergic receptor [α and β adrenergic receptor subtypes were not described until the classic study of Ahlquist[19]]. For R(−)-adrenaline, these groups were proposed to be (a) the basic nitrogen common to all sympathomimetic amines, (b) the phenyl group whose binding to the receptor was proposed to be enhanced by *meta* and/or *para* phenolic hydroxyl groups, and (c) the benzylic hydroxyl group of the β-carbon atom. According to the Easson-Stedman Hypothesis as modified by Blaschko[11] and Beckett[12], these three groups are in a most favorable stereochemical configuration for interaction with adrenergic receptors for only the R(−)-isomer of adrenaline. As far as S(+)-adrenaline and its β-desoxy derivative (epinine) are concerned, the β-hydroxyl group is either incorrectly oriented or absent, and therefore not available for interaction with the receptor. Thus, only a two point attachment is considered possible for these isomers. This presumably would explain the lower activities of the S(+)-isomer and corresponding desoxy derivative of adrenaline relative to the R(−)-isomer, and would also account for the fact that the S(+)-isomer and corresponding β-desoxy derivative are equal in activity to each other. The Easson-Stedman Hypothesis, as illustrated schematically for noradrenaline in Fig. 2, has also been shown by Patil *et al.*[20,21] to apply to a large series of sympathomimetic phenethylamines. However, this extension is only valid when *indirect* sympathomimetic activity, which is marked for the desoxy derivatives relative to the S(+)-isomers[22,23], is eliminated and only *direct* postsynaptic effects are considered. Further studies have shown that the Easson-Stedman Hypothesis holds true for all four known adrenergic receptor subtypes (α_1, α_2, β_1, β_2)[5,13,14,20,21].

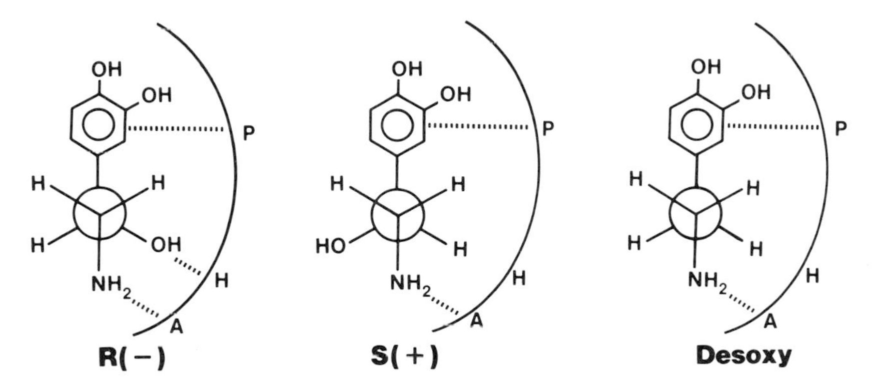

Figure 2. *Schematic representation of the Easson-Stedman Hypothesis for interaction of the R(−)- and S(+)-isomers of noradrenaline, and the corresponding β-desoxy derivative, dopamine, with adrenergic receptors. P, H, and A represent three hypothetical binding sites to which attach the phenyl, hydroxyl and amino functional groups of phenethylamines, respectively.*

The significance of the Easson-Stedman Hypothesis in relation to the imidazolines is not as clear as for the β-phenethylamines. When imidazolines structurally related to tolazoline are hydroxyl substituted at a position analogous to the β-carbon atom of the phenethylamines, α_1- and α_2-adrenergic activities are either unchanged or markedly *decreased*[24-28]. According to the Easson-Stedman Hypothesis, one would have expected α-adrenergic activity to *increase* if the hydroxyl group, when oriented in the proper configuration, were critical to the attachment of the agonist to the α-adrenergic receptor. In addition, the difference in activity between the R(−)- and S(+)-isomers of imidazolines with an asymmetric center at a position analogous to the β-carbon atom of phenethylamines is small[18,25–28], in contrast to the much larger differences of two to three orders of magnitude observed for the phenethylamines[14,21,29]. These results have prompted the suggestion that the α-adrenergic effects of the imidazolines do *not* adhere to the Easson-Stedman Hypothesis[24–28,30] while the phenethylamines most definitely do[14,21]. In spite of the fact that the α-adrenergic effects of the imidazolines do not conform to the Easson-Stedman Hypothesis, some preliminary findings indicate that the weak β-adrenergic effects of certain optically active 'catechol-imidazolines' *may* adhere to the Easson-Stedman Hypothesis[30].

In addition, it has been proposed that while the phenethylamines interact with the adrenergic receptors by a three-point attachment[10,14], the imidazolines may interact with α-adrenergic receptors by only a two-point attachment[18]. The possibility must be considered that the phenethylamines and imidazolines interact differently with α-adrenergic receptors[25,28,31–34].

b. Phenethylamines with asymmetry at the α-carbon atom. The stereoisomers of α-methyldopamine are commonly employed to study the stereochemical demands made by adrenergic receptors for phenethylamines with asymmetry at the α-carbon atom. Patil and Jacobowitz[35] have shown that 2S(+)-α-methyldopamine is more active than the corresponding 2R(−)-enantiomer in trachea from reserpine pretreated guinea pigs, a tissue containing predominantly β_2-adrenergic receptors. The effects of both isomers were antagonized by propranolol. Likewise, the β_1-adrenergic receptor mediated inotropic effects of α-methyldopamine are also confined to the 2S(+)-isomer[36]. Furthermore, in the vas deferens from the reserpine pretreated rat, it was observed that both isomers of α-methyldopamine were extremely weak agonists at α_1-adrenergic receptors, with little or no difference between the isomers being detected[35]. In agreement with these findings are those of Ruffolo and Waddell[6] who found the enantiomers of α-methyldopamine to be equiactive at α_1-adrenergic receptors in aorta from reserpine pretreated guinea pigs. In contrast to the equal potencies of the isomers of α-methyldopamine at α_1-adrenergic receptors is the marked preference shown by α_2-adrenergic receptors in field-stimulated guinea pig ileum for the 2S(+)-isomer over the enantiomeric 2R(−)-isomer[6].

In addition to the high stereoselectivity shown by the α_2-adrenergic receptor, but not by α_1, for the isomers of α-methyldopamine, marked differences also exist between these isomers in their α_2/α_1 selectivities. Thus, while 2R(−)-α-methyldopamine showed only a two-fold preference for α_2-adrenergic receptors over α_1, its enantiomer, 2S(+)-α-methyldopamine, displayed a 23-fold preference for the α_2-receptor[6].

According to the Easson-Stedman Hypothesis, dopamine, which lacks the β-hydroxyl group, will attach to both the α_1-[14,20,21] and α_2-[5] adrenergic receptors by only a two point attachment involving the catechol and aliphatic nitrogen atom. Since the two stereoisomers of α-methyldopamine are equipotent with dopamine at α_1-*adrenergic receptors*[6], it is concluded that both isomers of α-methyldopamine likewise bind to the α_1-receptor by only two points, and that this receptor does not interact significantly with the substituent at the α-carbon atom. Conversely, the 2S(+)-isomer of α-methyldopamine is significantly more potent at the α_2-adrenergic receptor than either the 2R(−)-isomer or dopamine (desmethyl analog), with the two latter compounds being equally active to each other. These results indicate that the α_2-adrenergic receptor, in marked contrast to the α_1, has the ability to interact with the α-methyl group of α-methyldopamine when it is present and in the *optimal 2S stereochemical configuration.* Since dopamine would appear to bind to the α_2-adrenergic receptor by only two points of attachment [*i.e.*, Easson-Stedman Hypothesis is also valid for α_2-receptors[5]], it follows that 2R(−)-α-methyldopamine likewise binds to the α_2-adrenergic receptor by only two points since this isomer is equipotent with dopamine at this receptor. However, the 2S(+)-isomer of α-methyldopamine, which is more potent than its enantiomer or dopamine, may bind to the α_2-adrenergic receptor by a three-point mode of attachment involving the catechol, nitrogen and α-methyl group[6]. This hypothesis involving asymmetry at the α-carbon atom of phenethylamines is presented schematically in Fig. 3 as it would differentially apply to α_1- and α_2-adrenergic receptors. The model calls for an additional recognition site existing only on the α_2-adrenergic receptor which can interact with and/or accommodate the α-methyl group of phenethylamines so substituted (*i.e.*, α-methyldopamine, α-methylnoradrenaline, etc.). 2R(−)-α-Methyldopamine, which possesses the α-methyl group but in the incorrect orientation, and the desmethyl derivative (*i.e.*, dopamine) which does not possess the α-methyl group, are predicted to be less active than the 2S(+)-isomer at α_2-adrenergic receptors (but not α_1-receptors) presumably because these two compounds may only bind to the α_2-adrenergic receptor by a two-point attachment.

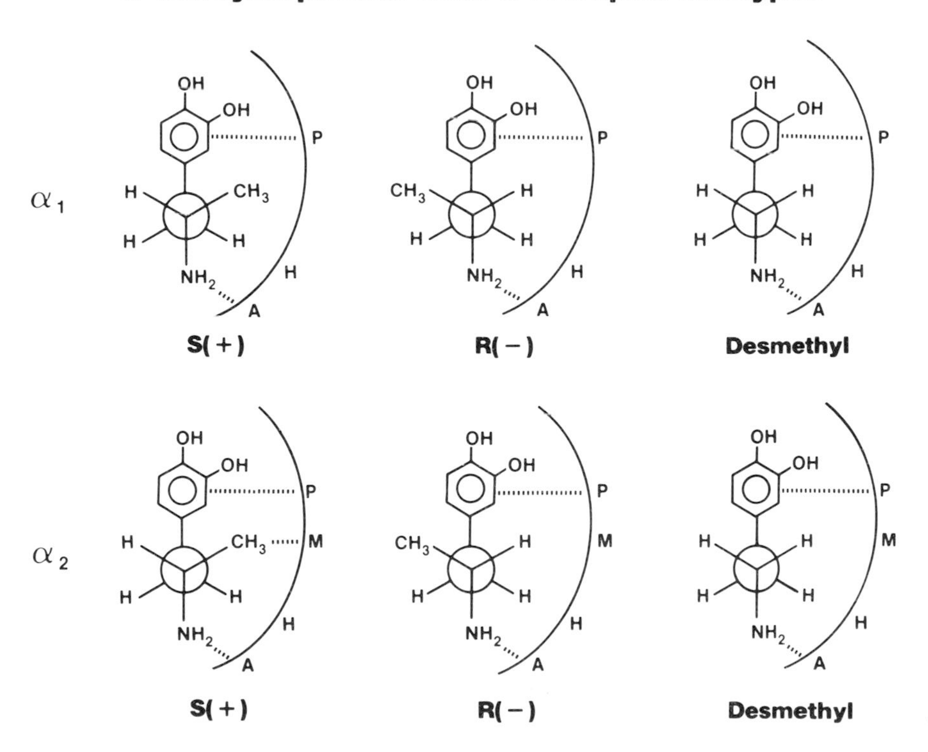

Figure 3. *Schematic representation of the possible interaction of α-methyl substituted phenethylamines with α_1- and α_2-adrenergic receptors. The hypothetical binding sites, P, H, and A are the same as indicated in Fig. 2, in addition to which the site M is proposed to exist only on α_2-adrenergic receptors to accomodate the α-methyl substituent.*

c. Phenethylamines with asymmetry at both the α- and β-carbon atoms. The stereochemical requirements made by adrenergic receptors for phenethylamines with two asymmetric centers are more complicated than those discussed above since four diastereomers (*i.e.*, two enantiomeric pairs) exist. The four possible isomers of α-methylnoradrenaline are shown in Fig. 4. The adrenergic receptors are strict in their configurational requirements for phenethylamine agonists with two asymmetric centers. For example, Patil and Jacobowitz[35] have established that the β_2-adrenergic receptor of guinea pig trachea is highly selective for only the 1R-2S(-)-*erythro*-isomer of α-methylnoradrenaline, with the remaining three isomers being inactive. Goldberg *et al.*[7], employing radioligand binding techniques, have recently shown that the same stereochemical requirements apply to the β_1-adrenergic receptor in rat brain. Likewise, the α_1-[35] and α_2-[5,7] adrenergic receptors are also highly selective for the 1R,2S(-)-*erythro* isomer of α-methylnoradrenaline.

α-Methylnoradrenaline Isomers

1R,2S(−)	1S,2R(+)	1R,2R	1S,2S
erythro	*erythro*	*threo*	*threo*

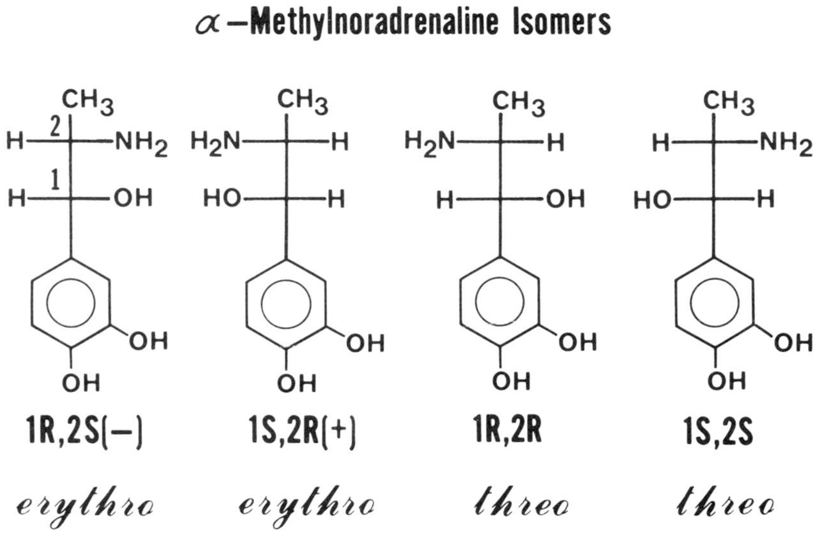

Figure 4. *The four diastereoisomers of α-methylnoradrenaline presented in the Fischer projection.*

Comparison of the stereochemical requirements of α_1- and α_2-adrenergic receptors illustrate several important differences between these two α-receptor subtypes. The isomeric activity ratio for the stereoisomers of noradrenaline was reported by Ruffolo *et al.*[5] to be approximately 107-fold for α_1-adrenergic receptors and 479-fold for α_2-adrenergic receptors. These findings suggest that quantitative differences exist in the configurational demands made by α_1- and α_2-adrenergic receptors, with the demands made by α_2-receptors being more stringent than those made by α_1-receptors, at least in the test systems employed. The case for the isomers of α-methylnoradrenaline was even more illustrative of these differences. For the enantiomeric 1R,2S(-)-*erythro* and 1S,2R(+)-*erythro* isomers of α-methylnoradrenaline, an isomeric activity ratio of 60-fold existed for α_1-adrenergic receptors in guinea pig aorta, whereas the ratio for α_2-adrenergic receptors in field-stimulated guinea pig ileum was 550-fold[5]. It was concluded, therefore, that the stereochemical demands made by α_2-adrenergic receptors are more stringent than those made by α_1-receptors, especially for phenethylamines possessing two asymmetric centers[5].

In the previous section it was argued that the α_2-adrenergic receptor could recognize and/or bind α-methyl substituents of phenethylamine agonists (see Section 1b.). Conversely, the α_1-adrenergic receptor did not have this ability. This hypothesis may be extended to include the interactions of α-methylnoradrenaline (two asymmetric centers) with α_1- and α_2-adrenergic receptors, and explain possible differences between α-methylnoradrenaline and noradrenaline. The most active isomers of α-methylnoradrenaline and noradrenaline at α_1- and α_2-adrenergic receptors are those in which the β-hydroxyl group is in the R absolute configuration[5,7,21]. Previous investigations which have shown that the Easson-Stedman Hypothesis is valid for both α-receptor subtypes[5] indicate that a three-point attachment is likely for 1R(–)-noradrenaline interacting with both α_1- and α_2-adrenergic receptors. Since the α_1-effects of 1R(–)-noradrenaline are not increased, or may even be *decreased*[5,18,21,37–39] by α-methyl substitution, it is concluded that the 1R,2S(–)-*erythro*-isomer of α-methylnoradrenaline also binds by a three-point attachment to α_1-adrenergic receptors, consistent with the hypothesis presented above which suggests that the α_1-receptor lacks the ability to interact with the α-methyl substituent. Conversely, addition of an α-methyl group to noradrenaline significantly enhances activity at α_2-adrenergic receptors[5–7,37] indicating that this receptor subtype can recognize the α-substituent *when oriented correctly* in the S absolute configuration. Since it appears that the α_2-adrenergic receptor can also recognize the catechol, β-hydroxyl and amino groups [*i.e.,* Easson-Stedman Hypothesis is valid for α_2[5]], it is proposed that the 1R,2S(–)-*erythro*-isomer of α-methylnoradrenaline binds to the α_2-adrenergic receptor by a *four*-point mode of attachment (*i.e.,* catechol, β-hydroxyl, amino and α-methyl groups) as opposed to the α_1-adrenergic receptor where only three groups may be involved in the binding of this same compound. The proposed interactions of 1R(–)-noradrenaline and 1R,2S(–)-*erythro*-α-methylnoradrenaline with both α-adrenergic receptor subtypes are illustrated schematically in Fig. 5. Note the proposed three-point interaction of 1R(–)-noradrenaline and 1R,2S(–)-*erythro*-α-methylnoradrenaline with α_1-adrenergic receptors. At α_2-adrenergic receptors, 1R(–)-noradrenaline will still interact by only a three-point attachment whereas 1R,2S(–)-*erythro*-α-methylnoradrenaline is proposed to interact with the α_2-adrenergic receptor by a four-point attachment which includes the α-methyl substituent. This model is consistent with the observations that 1R(–)-noradrenaline is equipotent with, or slightly more potent than, 1R,2S(–)-*erythro*-α-methylnoradrenaline at α_1-adrenergic receptors[5,6], whereas 1R,2S(–)-*erythro*-α-methylnoradrenaline is significantly more potent than 1R(–)-noradrenaline at α_2-adrenergic receptors[5,37]. An alternative hypothesis involving slightly different conformations of noradrenaline and α-methylnoradrenaline has been offered by Triggle[16].

d. Asymmetry elsewhere in phenethylamines. Simple phenethylamines such as noradrenaline, adrenaline and α-methylnoradrenaline have only two carbon atoms separating the aromatic ring and aliphatic nitrogen atom. For these simple phenethylamines, up to two points of asymmetry may exist; at the α- and β-carbon atoms. However, several phenethylamines with relatively large N-substituents have been synthesized, and many of these compounds possess an additional asymmetric center. Few such compounds have been resolved into their component stereoisomers and evaluated for activities in adrenergic test systems to assess configurational demands about these different asymmetric centers. Dobutamine is one compound, however, possessing an unusual point of asymmetry in which the individual stereoisomers have been evaluated in detail with some striking results. Dobutamine (Fig. 6) is an inotropic agent capable of increasing myocardial contractility at doses that have little or no effect on heart rate[40–43] or blood pressure[43,44]. The point of asymmetry exists on the rather bulky N-substituent. While it is generally observed

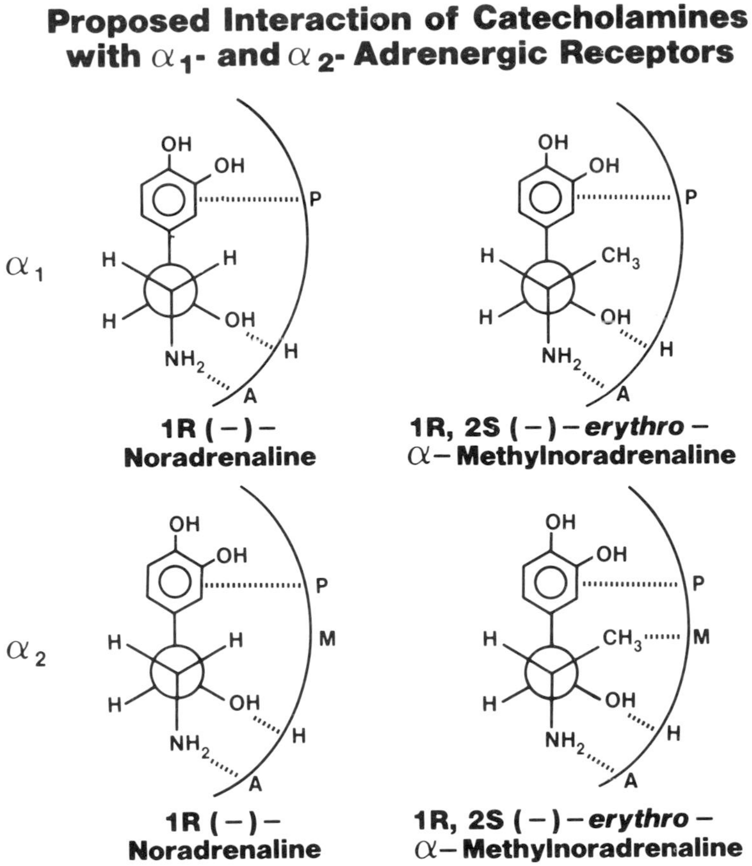

Proposed Interaction of Catecholamines with α_1- and α_2- Adrenergic Receptors

α_1

1R (−)- Noradrenaline | **1R, 2S (−)-*erythro*- α-Methylnoradrenaline**

α_2

1R (−)- Noradrenaline | **1R, 2S (−)-*erythro*- α-Methylnoradrenaline**

Figure 5. *Possible interactions of noradrenaline and α-methylnoradrenaline with α_1- and α_2-adrenergic receptors. The proposed sites of interaction, P, H, M, and A are the same as in Fig. 3.*

that the individual stereoisomers of phenethylamines possess qualitatively similar pharmacological activities and differ mainly in potency[14], dobutamine is unusual in that the individual stereoisomers display marked differences in their overall pharmacological profiles as well as large differences in potency. *In vitro* studies of the individual stereoisomers of dobutamine[45] indicate that the (-)-isomer is a potent α-adrenergic agonist and a weaker β-agonist relative to the (+)-isomer which is a strong β-agonist and possesses no α-agonist activity. In fact, (+)-dobutamine has been found to be an α-receptor antagonist capable of blocking the potent α-agonist effects of its enantiomer. These results have been subsequently confirmed *in vivo*[46]. In the pithed rat, the (-)-isomer of dobutamine is a potent directly-acting pressor agent whereas the (+)-isomer is only a weak pressor agent. The pressor activities of both stereoisomers of dobutamine are mediated exclusively by postsynaptic vascular α_1-adrenergic receptors.

Both stereoisomers of dobutamine lower blood pressure in reserpine and phenoxybenzamine pretreated pithed rats with vascular tone induced by a constant infusion of angiotensin II. However, in contrast to the pressor effects of the stereoisomers of dobutamine, this β_2-adrenergic receptor mediated vasodepressor[47] effect was greatest for the (+)-isomer[46]. Likewise, both isomers mediate a positive

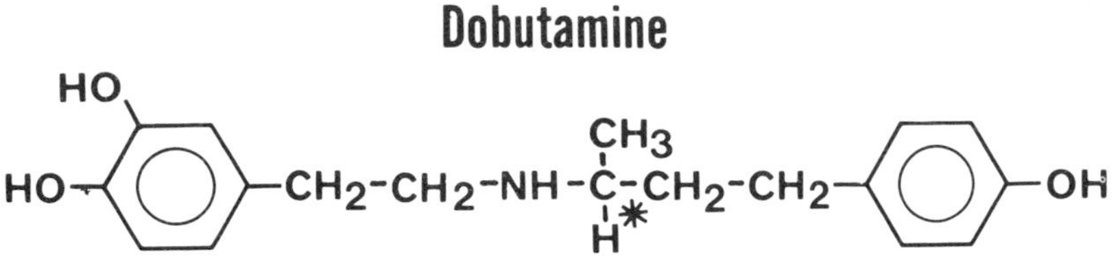

Dobutamine

HO
HO–⟨benzene⟩–CH_2-CH_2-NH-C(CH$_3$)(H)*-CH_2-CH_2–⟨benzene⟩–OH

Figure 6. *Chemical structure of the inotropic agent, dobutamine. The asterick denotes the point of asymmetry.*

chronotropic response in reserpine and phentolamine-pretreated pithed rats, and this β_1 effect[47] is also greatest for the (+)-isomer[46]. Thus, the (-)-isomer of dobutamine is a relatively potent α_1-adrenergic agonist and a relatively weaker β_1- and β_2-agonist, whereas its enantiomer, (+)-dobutamine, is a strong β_1 and β_2-agonist and possesses only minimal α_1-agonist activity.

e. Asymmetry in imidazolines. Optically active centers in imidazoline agonists are rare. However, several examples of optically active imidazolines are known and provide some insight into how this unique class of α-adrenergic agonists interacts with α-adrenergic receptors. As stated earlier, the Easson-Stedman Hypothesis predicts the following order of potency of phenethylamines possessing an asymmetric hydroxyl-substituted benzylic carbon atom: R(-) > S(+) = desoxy. The Easson-Stedman Hypothesis has been shown to be valid for a large series of phenethylamines[14], but does not appear to apply to the imidazolines. The imidazolines in Fig. 7 possessing asymmetric carbon atoms at positions analogous to the β-carbon of phenethylamines have been synthesized. Ruffolo *et al.*[24,25] have shown that hydroxyl substitution of compound **1** (desoxy) to yield compound **2** [($\pm$)-racemate] results in a 4 to 10-fold *decrease* in activity as opposed to a two order of magnitude *increase* in activity predicted by the Easson-Stedman Hypothesis. Hydroxyl substitution of compound **3** (desoxy) to yield the R(-)- and S(+)-isomers of compound **4** have also been studied. In a variety of α_1-adrenergic test systems, the rank order of potency for these isomers was as follows: desoxy $\geqslant$ R(-) > S(+)[26,28,30,48,49]. This order of potency is clearly different than that predicted by the Easson-Stedman Hypothesis. At α_2-adrenergic receptors, the R(-)-isomer was found to be only 6-fold more potent than the S(+)-isomer; however, the corresponding desoxy derivative was found to be a very potent partial agonist whose dose-response curve was positioned to the left of the R(-)-isomer[26]. Thus, it may be concluded that the Easson-Stedman Hypothesis does not apply to either the α_1- or α_2-adrenergic effects of imidazoline agonists in spite of the fact that it does predict the α-adrenergic effects of the phenethylamines.

One point of asymmetry also exists in tetrahydrozoline (Fig. 8). It has been demonstrated *in vitro*[25,50] and *in vivo*[51] that the activity of tetrahydrozoline resides predominantly in the (-)-isomer. The isomeric activity ratio for the enantiomers of tetrahydrozoline is less than 10-fold[25,50].

Optically active imidazolines with asymmetry existing at the 4 position of the imidazoline ring have also been synthesized (Fig. 8). The two major substituents that have been placed at this position are methyl and benzyl. Both substitutions decrease intrinsic activity and thereby change potent agonists into antagonists[52,53]. Virtually no stereoselectivity exists between the enantiomers of either the methyl or benzyl substituted imidazolines[50,52,53].

It is appropriate to compare the differences in the *degrees* of stereoselectivity observed between the imidazolines and phenethylamines. Isomeric activity ratios of

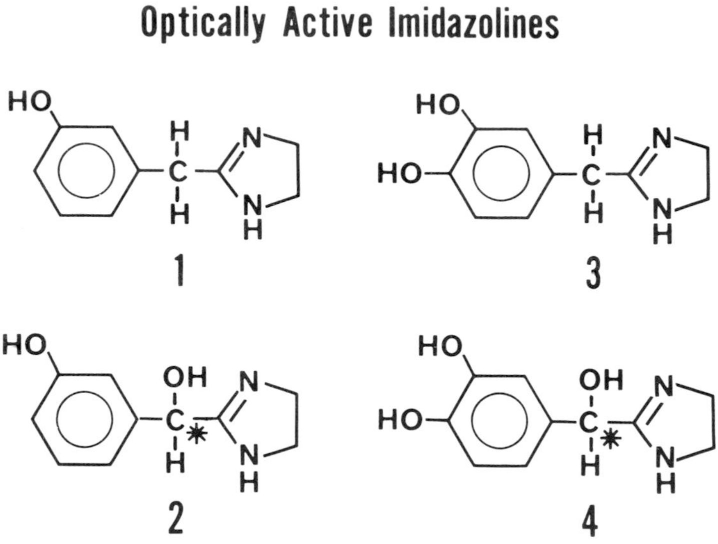

Optically Active Imidazolines

Figure 7. *Chemical structures of optically active imidazolines possessing benzylic hydroxyl groups, and their corresponding desoxy derivatives. The asterick denotes the point of asymmetry.*

Optically Active Imidazolines

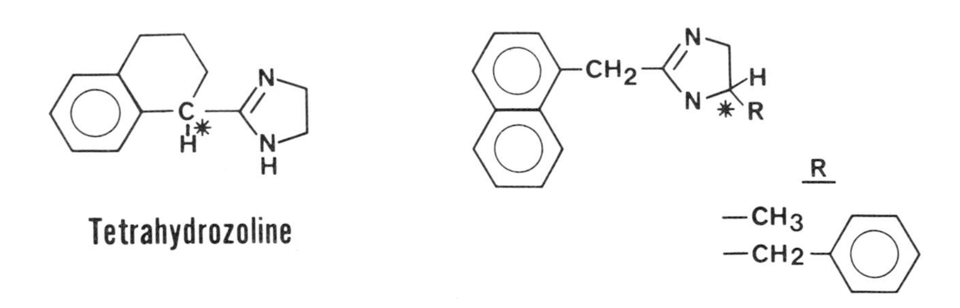

Figure 8. *Chemical structures of tetrahydrozoline, which is asymmetric at the benzylic carbon atom, and derivatives of naphazoline with points of asymmetry on the imidazoline ring.*

phenethylamines with asymmetry at the β-carbon atom are typically in excess of 100-fold. Asymmetry at the α-carbon of phenethylamines may be as great as 10-fold, and enantiomers of those phenethylamines with two points of asymmetry may show isomeric activity differences in excess of 500-fold. Conversely, isomeric activity differences for the imidazolines, *when they occur,* are typically low and rarely in excess of 5 to 10-fold. It would appear, therefore, that steric demands made by adrenergic receptors for phenethylamines are generally more stringent than those for the imidazolines.

2. Conformational requirements for directly-acting adrenergic agonists.

a. Phenethylamines. The most active enantiomer of noradrenaline and other phenethylamines at α- and β-adrenergic receptors is the R(−)-isomer[5,13, [illegible]]

When two asymmetric centers exist on a phenethylamine, as in α-methylnoradrenaline or ephedrine, the 1R,2S-(–)-isomer [*i.e.*, (–)-*erythro* form] is most active[5,6,13,14,18,21,35,54,55]. The *relative positions in space* of the three important functional groups (*i.e.*, phenyl, β-hydroxyl and aliphatic nitrogen) of a sympathomimetic amine when bound to α-adrenergic receptors is obtained from an analysis of the *conformational* demands made by the receptors. The exact conformation of phenethylamines required for interaction with adrenergic receptors is not known. However, consideration of the physical properties of these agonists in the solid state and in solution has allowed the energetically preferred conformations to be determined. Theoretical calculations[56–62] indicate that the preferred conformation of R(–)-noradrenaline in solution is the extended-*trans* conformation in which the amino and phenyl groups are at a dihedral angle of 180°. This conformation represents an energy minimum and hence greater stability and a greater probability of existence at any point in time. This conformation also appears to be stabilized by an intramolecular electrostatic or hydrogen bonding interaction between the amino and β-hydroxyl groups[61,62]. The 1R,2S-α-methyl,β-hydroxyl dissubstituted phenethylamines such as (–)-*erythro*-ephedrine and (–)-*erythro*-α-methylnoradrenaline also prefer this same extended-*trans* conformation[17,56,61,62]. X-Ray crystallographic studies of (–)-noradrenaline and (–)-ephedrine in the solid state[63,64] indicate that in this state, as in solution, the preferred conformation is the extended-*trans* form.

These studies regarding the preferred conformations of sympathomimetic amines in solution and in the solid state have led to speculation concerning which conformation of a sympathomimetic amine is required for binding to, and activation of, adrenergic receptors. However, the preferred conformation of an agonist in solution or in the solid state is not necessarily the same conformation required for interaction with the receptor[14,65]. Also, the preferred conformation of sympathomimetic amines in the relatively lipoidal region of the receptor, commonly called the 'biophase'[66,67], or near the 'active' site of the receptor where the physical environment may not resemble an aqueous solution, may differ from that conformation found to predominate in solution[65,68]

The molecular conformation of various biogenic amines has been reviewed in detail by Carlstrom *et al.*[64]. These authors propose, based on a large volume of literature dealing with the conformations of biogenic amines, that potent directly-acting sympathomimetic amines should have the five following characteristics: 1) a six membered aromatic ring system, 2) an extended ethylamine side chain oriented approximately perpendicular to the aromatic ring system, 3) a positively charged nitrogen (at physiological pH) on the ethylamine side chain, 4) a hydrophilic and hydrophobic side of the molecule resulting from the β-hydroxyl group being oriented on the same side of the molecule [*cis*] as the *meta* phenolic hydroxyl group of the aromatic ring, and 5) an R absolute configuration at the β-carbon atom. According to Carlstrom *et al.*[64], the amino, phenyl and β-hydroxyl groups, which Easson and Stedman[10] suggest are necessary for interaction with adrenergic receptors, will be in the appropriate configuration to interact with receptors when these five requirements are met as in the levorotatory isomers. For the dextrorotatory isomers, where the β-hydroxyl group is on the opposite side from the *meta* phenolic hydroxyl group, or for the desoxy derivative in which the β-hydroxyl group is absent, weaker activity should result (see Easson-Stedman Hypothesis, section 1a.).

Several attempts have been made to establish which of the many possible conformations of sympathomimetic amines is required for interaction with the α-adrenergic receptor subtypes. However, relatively little has been done to establish the conformational requirements made by the β-adrenergic receptor subtypes. The use of conformationally rigid analogs of noradrenaline has been suggested by Smissman and Gastrock[69]. These investigators have synthesized a series of conformationally restrained noradrenaline analogs that are derivatives of *trans* decalin for the purpose of establishing the conformational requirements made by adrenergic receptors for phenethylamines. However, the pharmacology of these conformationally restrained noradrenaline analogs has never been evaluated in detail due to the low agonist activity which results from the additional bulk used to restrict conformation. Hence, these efforts have not yielded the important information originally hoped for. However, Erhardt *et al.*[70], using two conformationally restricted analogs of dopamine (desoxy-noradrenaline), have attempted to establish the conformational demands made by the α_1-adrenergic receptor with molecules that possess relatively little unnecessary bulk. These investigators evaluated the α_1-adrenergic effects of the *trans*-extended and *cis*-folded isomers of 2-(3,4-dihydroxyphenyl)cyclopropylamine (Fig. 9) in rabbit oarta. The *trans*-extended form was found to be 5-fold more potent than the *cis*-folded analog, strongly suggesting that the *trans*-extended conformation, which is the highly preferred conformation in solution and in the solid state, is also that conformation preferred by the α_1-adrenergic receptor. Likewise, Ruffolo *et al.*[4] have investigated the conformational demands made by α_2-adrenergic receptors by evaluating the *trans*-extended and *cis*-folded analogs of 2-(3,4-dihydroxyphenyl)cyclobutylamine (Fig. 9) in field-stimulated guinea pig ileum. These results indicated that the presynaptic α_2-adrenergic receptor also preferred phenethylamines in the *trans*-extended conformation over the *cis*-folded form. It is concluded, therefore, that α_1- and α_2-adrenergic receptors have similar conformational requirements for activation by phenethylamine agonists.

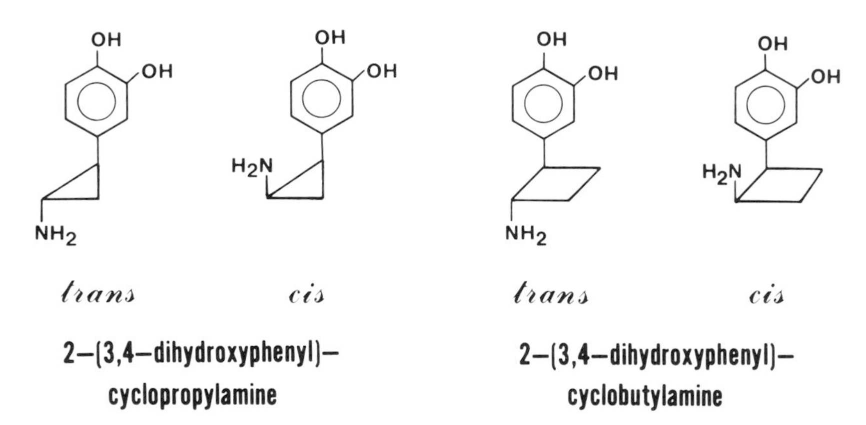

Figure 9. *Conformationally restricted cyclopropylamine and cyclobutylamine derivatives of dopamine used to establish the conformational requirements of* α_1- *and* α_2-*adrenergic receptors, respectively.*

b. Imidazolines. One of the first attempts to define the molecular conformation required for interaction of the imidazolines with α-adrenergic receptors was made by Pullman *et al.*[59]. In a quantum mechanical study of the conformational properties of naphazoline, these investigators concluded that the most stable conformation was one in which the naphthyl and imidazoline rings were mutually perpendicular with a dihedral angle of 90°. This conformation would place the aromatic ring and one of the imidazoline nitrogen atoms at a dihedral angle of approximately 180°, similar to what has been observed with the phenethylamines in the solid state and in solution. Consistent with this observation are reports that the free base of clonidine in solution prefers the conformation in which the phenyl and imidazoline rings assume a perpendicular arrangement[71-74]. X-Ray crystallographic studies of clonidine hydrochloride in the solid phase also show a nearly perpendicular arrangement between these two rings[75]. Although the perpendicular arrangement of the phenyl and imidazoline rings of clonidine has been attributed to steric forces around the relatively bulky *ortho* chlorine substituents[76,77], it has recently been suggested that even unsubstituted benzylimidazolines and phenyliminoimidazolines may also assume the same perpendicular arrangement of the phenyl and imidazoline rings in solution[73,74].

Although the preferred conformations of the imidazolines in solution and in the solid state have been established and shown to have resemblance to the phenethylamines, conformationally rigid imidazoline derivatives have not been studied in a highly quantitative manner to determine the conformational preference of the adrenergic receptor subtypes for imidazolines.

Stereochemical requirements for indirectly-acting adrenergic agonists

1. Indirectly-acting sympathomimetic amines and the role of neuronal uptake (uptake$_1$ or membranal uptake) and uptake into adrenergic storage vesicles (vesicular uptake).

It is now known that adrenergic agonists may be divided into two classes based on their mechanisms of action. The first class has been referred to as directly-acting sympathomimetic amines which have been previously discussed. These agonists are known to interact directly with adrenergic receptors, and a study of the configurational and conformational requirements for these compounds provides insight into the stereochemical preferences made by the postjunctional adrenergic receptors. The second class of adrenergic agonists is the indirectly-acting sympathomimetic amines such as tyramine and amphetamine. Agonists of this class do not interact directly with adrenergic receptors to a significant extent, but rather release endogenous (largely 'cytoplasmic') stores of neurotransmitter from sympathetic nerve terminals[78-80]. Before endogenous stores of catecholamine (generally noradrenaline in postganglionic sympathetic nerve terminals) can be released, the indirectly-acting sympathomimetic amine must first be transported into the nerve terminal by the 'cocaine-sensitive' amine uptake pump (uptake$_1$) located on the neuronal cell membrane[9]. Thus, a configurational and conformational analysis of predominantly indirectly-acting sympathomimetic amines will provide valuable information concerning the stereochemical demands made by uptake$_1$. It must be noted that many sympathomimetic amines act by *both* a direct and indirect nature[21], and care must be taken in evaluating the stereochemical selectivities of these 'mixed' adrenergic agonists.

While substantial quantities of neurotransmitter are known to be liberated from 'cytoplasmic' pools by indirectly-acting sympathomimetic amines, release of bound neurotransmitter from adrenergic storage vesicles located within the sympathetic nerve terminal has also been identified for some compounds (see ref. 80). Before vesicular release of neurotransmitter can occur, these indirectly-acting sympathomimetic amines must first be transported into the storage vesicle. Such compounds must, therefore, also be substrates for the vesicular transport system. Both reserpine-sensitive and reserpine-resistant vesicular uptake processes have been identified and it has been determined that certain indirectly-acting sympathomimetic amines release bound vesicular noradrenaline following transport into the storage vesicle *via* the reserpine-resistant system while others use the reserpine-sensitive carrier[80-83]. Since indirectly-acting sympathomimetic amines may liberate noradrenaline from the 'cytoplasmic' stores following uptake into the nerve terminal, and from vesicular stores following uptake into the storage vesicles, the configurational and conformational demands made by uptake$_1$ and the vesicular uptake mechanism(s) should be considered.

2. Configurational requirements for indirectly-acting phenethylamines.

a. Phenethylamines with asymmetry at the β-carbon atom. The ability of the neuronal uptake pump (uptake$_1$) to distinguish between optical isomers of phenethylamines with an asymmetric β-carbon atom has been highly disputed. Early reports indicated that the neuronal uptake pump, which transported indirectly-acting sympathomimetic amines into the sympathetic nerve terminal, was stereoselective for R(–)-isomers of phenethylamines[84-87]. However, the differences observed between enantiomers were rather small (2 to 5 fold) and not consistent from tissue-to-tissue and could not be observed in all tissues[88,89]. In fact, in certain tissues where neuronal uptake had previously been shown to be stereoselective for R(–)-isomers, this 'apparent' stereoselectivity was lost after inhibiting vesicular uptake which itself is a stereoselective process[90-92]. These results suggested that the stereoselectivity originally attributed to neuronal uptake may have in large part resulted from a stereoselectivity occurring at the level of the adrenergic storage vesicle. The question as to the absolute stereoselectivity of the neuronal membrane uptake pump for phenethylamines possessing asymmetry at the β-carbon atom has never been completely resolved, but Iversen *et al.*[93] have argued convincingly that uptake$_1$ is in all probability stereoselective for R(–)-isomers of β-hydroxyl substituted phenethylamines. The stereoselective differences displayed by the membrane uptake pump are at best only very small and would seem to vary from tissue-to-tissue, with some tissues being completely nonselective. The relatively small differences in stereochemical preference shown by the membrane uptake pump would seem to contrast with the relatively large stereochemical differences of two to three orders of magnitude shown by the various adrenergic receptors located postjunctionally.

While the question of stereoselectivity at the membrane uptake pump remains somewhat of an enigma, there is little doubt that the vesicular transport system for sympathomimetic amines is a highly stereoselective process. Von Euler and colleagues[94-97] have consistently demonstrated the existence of a stereoselective preference for R(–)-noradrenaline over the S(+)-enantiomer in storage vesicles isolated from bovine splenic nerves. Likewise, a similar stereochemical preference has been identified in isolated bovine chromaffin granules, or 'ghosts' prepared from them[98,99] and in storage vesicles from rat heart[100].

b. Phenethylamines with asymmetry at the α-carbon atom. Although the membranal neuronal uptake$_1$ pump displays little or no stereoselectivity for optical isomers with the point of asymmetry at the β-carbon atom, such is not the case when asymmetry occurs at the α-carbon atom where relatively large and reproducible stereoselectivity has been observed. Iversen[101] has reported a 20-fold difference in the abilities of 2S(+)- and 2R(–)-amphetamine to inhibit the neuronal uptake of noradrenaline, with the 2S(+)-isomer being the most potent. In addition, Marquardt *et al.*[102] investigated the ability of three phenethylamines with α-methyl substitutions to inhibit noradrenaline uptake in synaptosomes prepared from rat brain. For each of the three enantiomeric pairs, activity was approximately 3-fold greater for the 2S(+)-isomer compared to its corresponding 2R(–)-enantiomer.

Data are noticeably lacking concerning asymmetry at the α-carbon atom and stereoselectivity of vesicular uptake. However, inferences drawn from compounds with two asymmetric centers (see below) would tend to indicate that vesicular uptake would also favor the 2S(+)-isomer over the enantiomeric 2R(–)-isomer.

c. Phenethylamines with asymmetry at both the α- and β-carbon atoms. Patil and Jacobowitz[35] have determined by histochemical studies in iris from reserpine-pretreated rats that both the 1R,2S(-)-*erythro* and 1S,2R(+)-*erythro* isomers of α-methylnoradrenaline are substrates for neuronal uptake, with little or no stereochemical preference being observed for either isomer. While both *erythro*-isomers seemed to be transported by uptake$_1$ to similar degrees, it was observed that neither of the *threo*-isomers of α-methylnoradrenaline were potent substrates for uptake$_1$. Thus, while uptake$_1$ failed to distinguish between the *erythro* enantiomers, the membrane uptake pump could distinguish between the *erythro* and *threo* diastereomers. In spite of this apparent selectivity in rat iris, similar *rates* of neuronal uptake between the *erythro* and *threo* diastereomers have been reported in mouse and rabbit hearts, although the *erythro* isomers were selectively retained suggesting stereoselective vesicular uptake[103,104].

Several studies have shown that *only* the 1R,2S(-)-*erythro*-isomer of metaraminol can function as a false neurochemical transmitter[105,106]. For a compound to serve as a false neurotransmitter, it must be accumulated by the neuron and subsequently transported by the vesicular uptake pump[107]. Since all four isomers of metaraminol appear to be substrates for neuronal uptake, by inference, it would seem logical to assume that only the 1R,2S(-)-*erythro* isomer is subsequently accumulated and retained by the vesicular uptake pump, and that the storage vesicles display a marked stereoselectivity for only this one isomer. A more direct assessment of vesicular uptake of the stereoisomers of metaraminol has been made by Sugrue and Shore[83] who have demonstrated that only the 1R,2S(-)-*erythro*-isomer was a substrate for vesicular uptake. Consistent with this observation are the findings of Muscholl *et al.*[108] who demonstrated that while the initial rates of uptake of the isomers of α-methylnoradrenaline were the same, there was a selective retention of only the 1R,2S(-)-*erythro* form. The interpretation of these results was that the initial uptake by neuronal uptake$_1$ was not stereoselective for these phenethylamines with two asymmetric centers, whereas vesicular uptake and retention were highly stereoselective. Results consistent with this interpretation have been obtained by others[109,110].

The functional significance of these studies on the stereoselectivity of neuronal and vesicular uptake of phenethylamines with two points of asymmetry has been addressed by Patil *et al.*[20,21,54,55] who investigated the adrenergic effects of the four isomers of ephedrine. In the vas deferens, the activities of the isomers of ephedrine are *predominantly* indirect[21]. In this tissue, the indirect activity of ephedrine resides in only the 1R,2S(-)-*erythro*-isomer, with the three remaining isomers being only weakly active or inactive. It cannot be stated at the present time whether the stereoselectivity shown for the indirect effects of ephedrine result from stereoselectivity at uptake$_1$, vesicular uptake, or both, or whether, in fact, the stereoselectivity occurs at some other process such as displacement of endogenous noradrenaline from one or more 'bound' sites.

3. Conformational requirements of uptake$_1$ for indirectly-acting phenethylamines.

It has been proposed that the preferred conformation of noradrenaline for neuronal uptake is the *trans*-extended conformation in which the phenyl ring and aliphatic nitrogen are at a dihedral angle of 180°[111-114]. The use of conformationally restricted phenethylamine derivatives has largely confirmed that the *trans*-extended conformation is highly preferred by the neuronal uptake pump. Horn and Snyder[115] and Tuomisto *et al.*[116] have shown that *trans*-2-phenylcyclopropylamine (Fig. 10) is two to three orders of magnitude more potent than the corresponding *cis*-folded form in inhibiting noradrenaline uptake in the central nervous system. Miller *et al.*[117] have obtained similar results in peripheral tissues. Other studies using rigid or semirigid phenethylamine analogs also indicate a high selectivity of the amine uptake pump for the *trans*-extended conformation[112,113,117–120]. Some evidence exists, however, to indicate that *gauche* conformations are not without activity[14,121]. The conformational demands made by the vesicular transport system for phenethylamines have not been defined.

2–Phenylcyclopropylamine Isomers

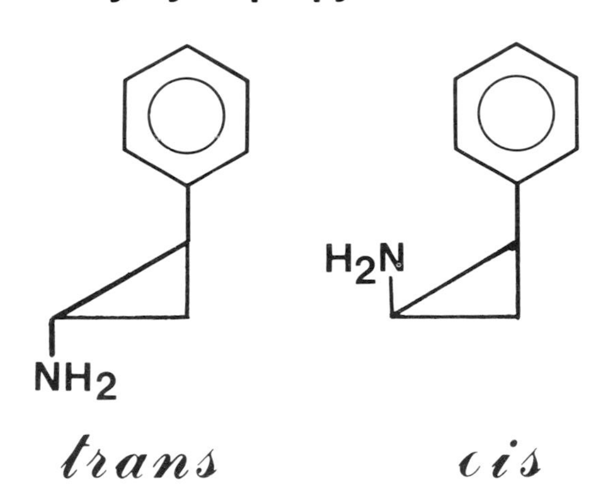

Figure 10. *Conformationally restricted cyclopropylamine derivatives used to establish the conformational requirements of* $uptake_1$.

Stereochemical requirements for adrenergic blocking agents

1. α-Adrenergic Antagonists.

Stereoselectivity in α-adrenergic blocking agents has been a relatively disappointing area of research. One reason for this is the fact that a vast majority of α-blocking agents do not possess a point of asymmetry. Several optically active imidazoline blocking agents have been prepared, but differences in activity between enantiomers is usually not observed[52,53].

One particularly interesting class of α-adrenergic blocking agents is the benzodioxans which do possess one or more points of asymmetry. Several benzodioxans have been synthesized and their absolute configurations determined by Nelson and coworkers[122,123]. Benzodioxans such as WB-4101, prosympal, and piperoxan (Fig. 11) possess one asymmetric center, and it has been determined that the S configuration is always more potent than the R[50,122–124]. Dibozane has two asymmetric centers (Fig. 11) and it has been determined that the S,S-isomer is more potent than the enantiomeric R,R-isomer[50]. The *meso* form (dibozane is a symmetrical molecule) is similar in potency to the S,S-isomer[50]. Nelson *et al.*[122] have argued convincingly that the conformational distribution of the aminoalkyl, oxygen and aromatic functional groups of the S-benzodioxans is similar to that of R(–)-adrenaline, thus accounting for the α-blocking activities and stereoselectivities of this class of compounds.

Recently, Weitzell *et al.*[125] have reported marked and surprising differences in the pre- and postsynaptic α-blocking activities of yohimbine isomers. Differences in potency as well as α-receptor subtype selectivity were reported and the authors proposed that several of the yohimbine isomers would be useful as pharmacological tools to subclassify α-adrenergic receptors. In a recent review, McGrath[126] has indicated that the potency of four isomers of yohimbine proceeds along the following or-

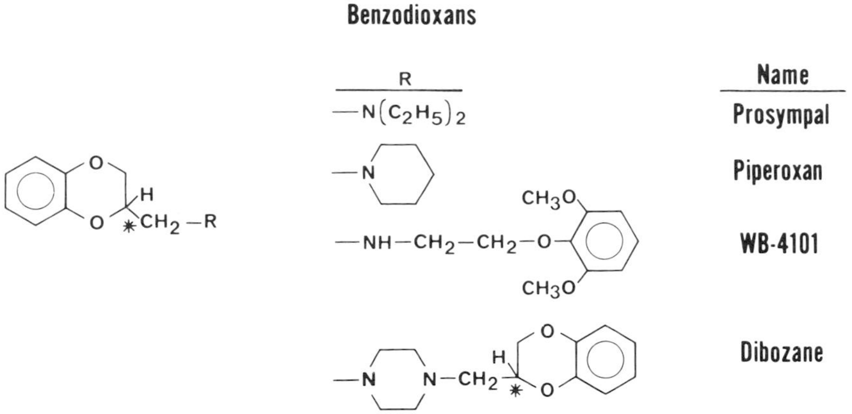

Benzodioxans

R	Name
$-N(C_2H_5)_2$	Prosympal
—N(piperidine)	Piperoxan
$-NH-CH_2-CH_2-O-$(2,6-dimethoxyphenyl)	WB-4101
—N(piperazine)N—CH_2—(benzodioxan)*	Dibozane

Figure 11. *Chemical structures of several optically active α-adrenergic blocking agents of the benzodioxan class. The asterick denotes the point of asymmetry.*

der at α_1-adrenergic receptors: apoyohimbine > corynanthine > yohimbine > rauwolscine. Potencies at α_2-adrenergic receptors were as follows: apoyohimbine > rauwolscine > yohimbine > corynanthine. In spite of these relative potencies at α_1- and α_2-adrenergic receptors, the selectivities of these compounds for α_2 over α_1 adrenergic receptors is: rauwolscine > yohimbine > apoyohimbine > corynanthine. No doubt these compounds will become increasingly important tools to probe α-adrenergic receptors.

The stereoisomers of the irreversible α-receptor blocking agent, phenoxybenzamine, have been prepared and evaluated by Portoghese *et al.*[127]. A 15-fold difference was observed in the *rate* at which these stereoisomers alkylated α-adrenergic receptors, with the (+)-isomer having the faster rate. Portoghese *et al.*[127] have argued that the intrinsic alkylating activities of the phenoxybenzamine isomers will be the same and that both isomeric forms will interact with the receptor through the highly reactive aziridinium species which will be formed from both enantiomers at the same rate. It was proposed, therefore, that the observed isomeric difference resulted from differences in affinities of the asymmetric precursors for the receptor. Other irreversible α-receptor blocking agents, such as N,N-dimethyl-β-chlorophenethylamine, have been shown not to be stereoselective[128,129], since both isomers alkylate the receptor through one common highly reactive symmetrical intermediate in which asymmetry is lost prior to alkylation.

2. β-Adrenergic Antagonists.

Unlike α-adrenergic blocking agents, β-adrenergic antagonists closely resemble β-agonists in both chemical structure and stereochemical requirements. Most β-adrenergic blockers possess a hydroxyl group in a position analogous to the β-hydroxyl group of isoproterenol. The most active isomers of β-blocking agents with asymmetry at this point have configurations identical to those of the most active isomers of β-agonists[14]. The similarity in stereochemical requirements for β-agonists and β-antagonists suggests that the latter may be more specific in their attachment to β-receptors than α-antagonists are in their attachment to α-receptors where little similarity exists between agonists and antagonists[14]. Methyl substitution of β-adrenergic antagonists at the α-carbon atom has been reported to decrease potency[62]. It would appear, however, that in these β-blocking agents where asymmetry exists at both the α- and β-carbon atoms, the most active isomer will have the same absolute configuration as 1R,2S(–)-*erythro*-α-methylnoradrenaline. Butedrine is a β-adrenergic blocking agent with two points of asymmetry that are remote from each other. At the 1 position is the hydroxyl group, and at the 3 position is the methyl substituent. All four isomers have been tested and both the 1R,3R- and 1R,3S-isomers are potent β-blocking agents with the 1R,3R-isomer perhaps being slightly more active[14]. The 1S,3S- and 1S,3R-isomers are relatively weak β-blocking agents presumably due to the incorrect stereochemistry at the 1 position.

3. 'Mixed' α- and β-Adrenergic Antagonists.

The most useful information concerning the stereoselectivity of adrenergic blocking agents will come from the relatively new class of blocking agents which block both α- and β-adrenergic receptors. Labetalol and medroxalol are representative members of this class of 'mixed' antagonists and each possesses two asymmetric centers (Fig. 12) with four possible isomers. Recent studies of labetalol[130-133], and

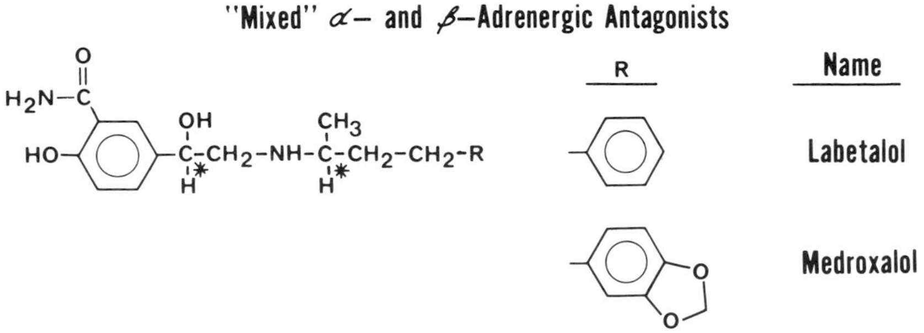

"Mixed" α– and β–Adrenergic Antagonists

Figure 12. *Chemical structures of two 'mixed' α- and β-adrenergic receptor antagonists, labetalol and medroxalol. Each compound has two points of asymmetry (asterick), and differential α- and β-adrenergic receptor blockade occurs among the individual isomers that comprise the unresolved mixture of diastereoisomers.*

medroxalol[134] have demonstrated that the α- and β-adrenergic blocking properties are not distributed uniformely among the individual stereoisomers. The four isomers of labetalol have now been resolved and their absolute configurations assigned. The α- and β-adrenergic blocking activities of these isomers have been evaluated in the dog by Brittain *et al.*[130] and the relative potencies listed in Table 1 have been calculated from their results. It is clear that the α_1-adrenergic blocking activity of labetalol resides predominantly in the S,R-isomer, whereas the β_1- and β_2-blocking effects occur predominantly in the R,R-isomer. That one of the four isomers

Table 1. *Relative α- and β-adrenergic blocking activities of the stereoisomers of labetalol in the dog (from ref. 130)*

	Relative Potencies		
Absolute Configuration	α_1	β_1	β_2
Labetalol (R,R; S,S; R,S; S,R)	1	1	1
R,R	0.15	**2.27**	**2.18**
S,S	0.39	0.03	<0.02
R,S	0.23	0.15	0.09
S,R	**1.74**	0.04	0.02

should be selective for α-adrenergic receptors while another would demonstrate a distinct preference for β-adrenergic receptors is indeed remarkable. An interesting comparison may be made between labetalol and dobutamine (section 1d.). Both compounds are close structural analogs, but dobutamine has only one asymmetric center and this is on the relatively bulky N-substituent, an asymmetric position also common to labetalol. With dobutamine it was observed that the (-)-isomer was predominantly an α-adrenergic agonist while the (+)-enantiomer was predominantly a β-agonist. Thus, it may be that this relatively unusual position of asymmetry can impart differential α- and β-adrenergic receptor selectivities upon component stereoisomers.

Summary and conclusions

The Easson-Stedman Hypothesis [*i.e.,* R(-) > S(+) = desoxy] is the most generally applicable theory regarding the direct sympathomimetic activity of phenethylamines possessing one asymmetric center. The theory proposes a three-point attachment to the receptor (*via* the nitrogen, β-hydroxyl and phenyl ring) for the R(-)-isomer, whereas the S(+)-enantiomer and corresponding β-desoxy derivative bind by a two-point attachment (through the amino and phenyl groups) accounting for their lower activities. The Easson-Stedman hypothesis is valid for phenethylamines interacting with all adrenergic receptor subtypes (*i.e.,* α_1, α_2, β_1, β_2), but does not hold for the α-adrenergic effects of the imidazolines. Phenethylamines with two asymmetric centers (*e.g.* α-methylnoradrenaline) have four stereoisomers and the activity resides in the 1R,2S(-)-*erythro* form for all adrenergic receptors. The α_1- and α_2-adrenergic receptors prefer phenethylamines in the *trans*-extended conformation.

Indirectly-acting sympathomimetic amines act through liberation of endogenous noradrenaline, and as such, a stereochemical analysis of these compounds provides information about the demands made by the amine uptake pump (uptake$_1$) and the adrenergic storage vesicles. The amine uptake pump displays, at best, only a slight stereochemical preference for R(-)-isomers of β-hydroxy substituted phenethylamines, while the adrenergic storage vesicles are highly stereoselective. Neuronal uptake does show a stereochemical preference for the 2S(+)-isomers of α-methyl substituted phenethylamines over the corresponding 2R(-)-isomers, and thus it follows that the major activity of indirectly-acting sympathomimetic amines with two asymmetric carbon atoms (*e.g.,* ephedrine) resides predominantly in the 1R,2S(-)-*erythro* form. The *trans*-extended conformation of a phenethylamine is preferred over the *cis*-folded conformation by uptake$_1$.

β-Adrenergic receptor antagonists are structurally similar to the agonists, and similar stereochemical requirements are expected and, in fact, observed. Competitive α-blocking agents of the benzodioxan class, as well as isomers of yohimbine, do show stereoselectivity. Most irreversible α-blocking agents (except phenoxybenzamine) display no stereoselectivity suggesting that both isomeric forms of such compounds may interact with the receptor through one common, highly reactive symmetrical intermediate. 'Mixed' α- and β-adrenergic blocking agents are now known, and most of these possess two points of asymmetry. Differences in the relative α- and β-adrenergic blocking potencies exist among the individual stereoisomers of these 'mixed' blockers. This relatively new class of antagonists with dual blocking activities will prove useful in distinguishing between the stereochemical demands made by adrenergic receptors.

Acknowledgements

I would like to thank Drs. J. Scott Hayes and Patrick J. Murphy for their helpful comments concerning this manuscript, and Ms. Elaine Gardner for her assistance in its preparation.

References

1. Timmermans, P.B.M.W.M. and Van Zwieten, P.A., J. Auton. Pharmacol. *1*, 171 (1981).
2. Langer, S.Z. Biochem. Pharmacol. *23*, 1793 (1974).
3. Stjarne, L., Acta Physiol. Scand. *90*, 286 (1974).
4. R.R. Ruffolo Jr., K.S. Anderson and D.D. Miller, Mol. Pharmacol. *21*, 259 (1982)
5. R.R. Ruffolo Jr., E.L. Yaden and J.E. Waddell, J. Pharmacol. Exp. Ther. *222*, 645 (1982).
6. R.R. Ruffolo Jr. and J.E. Waddell, Life Sci. *31*, 2999 (1982).
7. M.R. Goldberg, C.-S. Tung, R.D. Feldman, H.E. Smith, J.A. Oates and D.J. Robertson, Pharmacol. Exp. Ther. *220*, 552 (1982).
8. M.R. Goldberg, C.-S. Tung, M. Ring, J.A. Oates, J.F. Gerkens and D. Robertson, Clin. Exp. Hypertension *A4*, 595 (1982).
9. L.L. Iversen, Adv. Drug Res. *2*, 239 (1965).
10. L.H. Easson and E. Stedman, Biochem. J. *27*, 1257 (1933).
11. H. Blaschko, Proc. Roy Soc. B. *137*, 307 (1950).
12. A.H. Beckett, Fortschr. Arzneimittel-Forschung *1*, 455 (1959).
13. P.N. Patil, J.B. LaPidus and A. Tye, J. Pharm. Sci. *59*, 1205 (1970).
14. P.N. Patil, D.D. Miller and U. Trendelenburg, Pharmacol. Rev. *26*, 323 (1974).
15. P.N. Patil and J.B. LaPidus, Ergeb. Physiol. *66*, 213 (1972).
16. D.J. Triggle, In: *Chemical Pharmacology of the Synapse,* ed. by D.J. Triggle and C.R. Triggle, Academic Press, New York, pp. 233-430, 1976.
17. P.S. Portoghese, Annu. Rev. Pharmacol. *10*, 51 (1970).
18. R.R. Ruffolo Jr. In: *Adrenoceptors and Catecholamine Action,* ed. by G. Kunos, Wiley-Interscience, New York pp. 1-50, 1983.
19. R.P. Ahlquist, Am. J. Physiol, *153*, 586 (1948).
20. P.N. Patil, J.B. LaPidus and A. Tye, J. Pharmacol. Exp. Ther. *155*, 1 (1967).
21. P.N. Patil, J.B. LaPidus, D. Campbell and A. Tye, J. Pharmacol. Exp. Ther. *155*, 13 (1967).
22. P.N. Patil, D.G. Patel and A. Tye Arch. Int. Pharmacodyn. Ther. *182*, 32 (1969).
23. R.R. Ruffolo Jr., D.D. Miller and P.N. Patil, Biochem. Pharmacol. *23*, 399 (1976).
24. R.R. Ruffolo Jr., R.D. Dillard, E.L. Yaden and J.E. Waddel, J. Pharmacol. Exp. Ther. *211*, 74 (1979).
25. R.R. Ruffolo Jr., E.L. Yaden, J.E. Waddell and R.D. Dillard, J. Pharmacol. Exp. Ther. *214*, 535 (1980).
26. R.R. Ruffolo Jr. P.J. Rice, A. Hamada, D.D. Miller and P.N. Patil, Eur. J. Pharmacol. *86,* 471 (1983).
27. R.R. Ruffolo Jr. and J.E. Waddell, J. Pharmacol. Exp. Ther. *224,* 559 (1983).
28. D.D. Miller, A. Hamada, P.J. Rice and P.N. Patil, Abstracts of the American Chemical Society, Division of Medicinal Chemistry, Abstract No. 38, March, 1980.
29. P.N. Patil, D.G. Patel and R.D. Krell, J. Pharmacol. Exp. Ther. *176*, 622 (1971).
30. R.R. Ruffolo Jr., P.N. Patil and D.D. Miller, Naunyn-Schmiedeberg's Arch. Pharmacol. In Press, 1983.
31. R.R. Ruffolo Jr., B.S. Turowski and P.N. Patil, J. Pharm. Pharmacol. *29*, 378 (1977).
32. R.R. Ruffolo Jr., D.D. Miller and P.N. Patil, In: *Recent Advances in the Pharmacology of Adrenoceptors,* ed. by E. Szabadi, C.M. Bradshaw and P. Bevan, Elsevier/North Holland Biomedical Press, pp. 45-50, 1978.
33. W. Kobinger, C. Lillie and L. Pichler, Circ. Res. *46* (Suppl. I), I-21 (1980).
34. D.R. Mottram, Br. J. Pharmacol. *75* (Suppl.), 138P (1982).
35. P.N. Patil and D. Jacobowitz, J. Pharmacol. Exp. Ther. *161*, 279 (1968).
36. D. Palm, W. Langeneckert and P. Holtz, Naunyn-Schmiedeb. Arch. Pharmak. u. exp. Path. *258*, 128 (1967).
37. K. Starke, T. Endo and H.D. Taube, Naunyn-Schmiedeberg's Arch. Pharmacol. *291*, 55 (1975).

38. J.C. Besse and R.F. Furchgott, J. Pharmacol. Exp. Ther. *197*, 66 (1976).
39. L.T. Williams, D. Mullikin and R.J. Lefkowitz, J. Biol. Chem. *251*, 6915 (1976).
40. R.R. Tuttle and J. Mills, Circ. Res. *36*, 185 (1975).
41. R.R. Tuttle and J. Mills, U.S. Patent, 3,987,200, 1976.
42. E.H. Sonnenblick, W.H. Frishman and T.H. LeJemtel, N. Engl. J. Med. *300*, 17 (1979).
43. C.V. Leier, P.T. Heban, P. Huss, C.A. Bush and R.P. Lewis, In: Internationales Dobutamin Symposion, ed. by W. Bleifeld, R. Gattiker, W. Schaper and W. Brade, pp. 177-183, Urban and Schwarzenberg, Munich, 1980.
44. C.-S. Liang and W.B. Hood, J. Pharmacol. Exp. Ther. *211*, 698 (1979).
45. R.R. Ruffolo Jr., T.A. Spradlin, G.D. Pollock, J.E. Waddell and P.J. Murphy, J. Pharmacol. Exp. Ther. *219*, 447 (1981).
46,. R.R. Ruffolo Jr. and E.L. Yaden, J. Pharmacol. Exp. Ther. *224*, 46 (1983).
47. B. Wilffert, P.B.M.W.M. Timmermans and P.A. Van Zwieten, J. Pharmacol. Exp. Ther. *221*, 762 (1982).
48. D.D. Miller, P.N. Patil and D.R. Feller, Abstracts, American Chemical Society, Division of Medicinal Chemistry, Abst. 5, 1982.
49. P.J. Rice, A. Hamada, D.D. Miller and P.N. Patil, The Pharmacologist *24*, 222, 1982.
50. H. Fuder, W.L. Nelson, D.D. Miller and P.N. Patil, J. Pharmacol. Exp. Ther. *217*, 1 (1981).
51. T.O. Yellin, M.S. Katchen, S.R. Lavenhar and E.G. Nelson, Acta. Physiol. Scand. Special Suppl. 219 (1978).
52. D.D. Miller, F.-L. Hsu, R.R. Ruffolo Jr. and P.N. Patil, J. Med. Chem. *19*, 1382 (1976).
53. F.-L. Hsu, A. Hamada, M. Booher, H. Fuder, P.N. Patil, and D.D. Miller, J. Med. Chem. *23*, 1232 (1980).
54. P.N. Patil, A. Tye and J.B. LaPidus, J. Pharmacol. Exp. Ther. *148*, 158 (1965).
55. P.N. Patil, A. Tye and J.B. LaPidus, J. Pharmacol. Exp. Ther. *149*, 199 (1965).
56. L.B. Kier, J. Pharmacol. Exp. Ther. *164*, 75 (1968).
57. L.B. Kier, J. Pharm. Pharmacol. *21*, 93 (1969).
58. L.B. Kier and E.B. Truitt Jr., J. Pharmacol. Exp. Ther. *174*, 94 (1970).
59. B. Pullman, J.L. Coubeils, P.H. Couriere and J.P. Gervois, J. Med. Chem. *15*, 17 (1972).
60. L. Pedersen, R.E. Hoskins and H. Cable, J. Pharm. Pharmacol. *23*, 216 (1971).
61. R.R. Ison, P. Partington and G.C.K. Roberts, Mol. Pharmacol. *9*, 756 (1973).
62. P.S. Portoghese, J. Med. Chem. *10*, 1057 (1967).
63. D.C. Phillips, Acta Crystallogr. *7*, 159 (1954).
64. D. Carlstrom, R. Bergin and G. Falkenberg, Q. Rev. Biophys. *3*, 257 (1973).
65. J.B. Robinson, B. Belleau and B. Cox, J. Med. Chem. *12*, 848 (1969).
66. R.F. Furchgott, Pharmacol. Rev. *7*, 183 (1955).
67. R.F. Furchgott, Ann. Rev. Pharmacol. *4*, 21 (1964).
68., H.F. Ridley, S.S. Chatterjee, J.F. Moran and D.J. Triggle, J. Med. Chem. *12*, 931 (1969).
69. E.E. Smissman and W.H. Gastrock, J. Med. Chem. *11*, 860 (1968).
70. P.W. Erhardt, R.J. Gorczynski and W.G. Anderson, J. Med. Chem. *22*, 907 (1979).
71. B. Rouot, G. Leclerc and C.G. Wermuth, Chim. Ther. *5*, 545 (1973).
72. K.H. Pook, H. Stahle and H. Daniel, Chem. Ber. *107*, 2644 (1974).
73. L.M. Jackman and T. Jen, J. Am. Chem. Soc. *97*, 2811 (1975).
74. A.P. de Jong and H. Van Dam, J. Med. Chem. *23*, 889 (1980).
75. V. Cody and G.T. DeTitta, Mol. Struct. *9*, 33 (1979).
76. P.B.M.W.M. Timmermans, P.A. Van Zwieten,C.M. Meerman-van Benthem, K. van der Meer and J.J.C. Mulder, Arzneim. Forsch. *27*, 2266 (1977).
77. C.M. Meerman-van Benthem, K. van der Meer, J.J.C. Mulder, P.B.M.W.M. Timmermans, and P.A. van Zwieten, Mol. Pharmacol. *11*, 667 (1975).
78. E. Muscholl, Naunyn-Schmiedeberg's Arch. Pharmacol. *240*, 234 (1960).
79. A. Carlsson, 'Mechanism of Release of Biogenic Amines', Pergamon Press, New York, 331, 1966.
80. U. Trendelenburg, In: *Handbook of Experimental Pharmacology,* vol. 33, ed. by H. Blaschko and E. Muscholl, Springer-Verlag, Berlin, pp. 336-362, 1972.
81. P. Lundborg and R. Stitzel, Br. J. Pharmacol. *29*, 342 (1967).

82. R.E. Stitzel and P. Lundborg, Br. J. Pharmacol. *29*, 99 (1967).
83. M.F. Sugrue and P.A. Shore, J. Pharmacol. Exp. Ther. *177*, 389 (1971).
84. L.L. Iversen, Brit. J. Pharmacol. *21*, 523 (1963).
85. L.L. Iversen, The uptake and storage of noradrenaline in sympathetic nerves, pp. 253, Cambridge University Press, Cambridge, England, 1967.
86. I.J. Kopin and W. Bridgers, Life. Sci. *2*, 356 (1963).
87. R.P. Maickel, M.A. Beaven and B.B. Brodie, Life Sci. *2*, 953 (1963).
88. S.B. Ross and A.L. Renyi, Acta. Pharmacol. Toxicol. *21*, 226 (1964).
89. P.R. Draskoczy and U. Trendelenburg, J. Pharmacol. Exp. Ther. *159*, 66 (1968).
90. U. Trendelenburg and P.R. Draskoczy, J. Pharmacol. Exp. Ther. *175*, 521 (1970).
91. R.D. Krell and P.N. Patil, J. Pharmacol. Exp. Ther. *182*, 101 (1972).
92. R.D. Krell and P.N. Patil, J. Pharmacol. Exp. Ther. *182*, 273 (1972).
93. L.L. Iversen, B. Jarrott and M.A. Simmonds, Brit. J. Pharmacol. *43*, 845 (1971).
94. U.S. Euler v. and F. Lishajko, Acta Physiol. Scand. *60*, 217 (1964).
95. U.S. Euler v. and F. Lishajko, Int. J. Neuropharmacol. *4*, 273 (1965).
96. U.S. Euler v. and F. Lishajko, Acta Physiol. Scand. *71*, 151 (1967).
97. L. Stjarne and U.S. Euler v., J. Pharmacol. Exp. Ther. *150*, 335 (1965).
98. G. Taugner, Naunyn-Schmiedeberg's Arch. Pharmacol. *274*, 299 (1972).
99. J.H. Phillips: Biochem. J. *144*, 319 (1974).
100. T.C. Westfall, Acta. Physiol. Scand. *63*, 336 (1965).
101. L.L. Iversen, Br. Med. Bull. *29*, 130 (1973).
102. G.M. Marquardt, V. DiStefano and L.L. Ling, Biochem. Pharmacol. *27*, 1497 (1978).
103. E. Muscholl and R. Lindmar, Naunyn-Schmiedeberg's Arch. Pharmacol. *257*, 314 (1967).
104. R. Lindmar, E. Muscholl and K.H. Rahn, Eur. J. Pharmacol. *2*, 317 (1968).
105. N.F. Albertson, F.C. McKay, H.E. Lape, J.O. Hoppe, W.H. Selberis and A. Arnold, J. Med. Chem. *13*, 132 (1970).
106. M.L. Torchiana, C.C. Porter and C.A. Stone, Arch. Int. Pharmacodyn. Ther. *174*, 118 (1968).
107. E. Muscholl: In: *Handbook of Experimental Pharmacology,* vol. 33, ed. by H. Blaschko and E. Muscholl, Springer-Verlag, Berlin, pp. 618-660, 1972.
108. E. Muscholl, E.F. Drews and R. Lindmar, Naunyn-Schmiedeberg's Arch. Pharmak. Exp. Path. *260*, 180 (1968).
109. B. Waldeck, Eur. J. Pharmacol. *2*, 208 (1967).
110. A. Carlsson, J. Meisch and B. Waldeck, Eur. J. Pharmacol. *5*, 85 (1968).
111. R.A. Maxwell, E. Chaplin and S. Batmanglidj Eckhardt, J. Pharmacol. Exp. Ther. *166*, 320 (1969).
112. R.A. Maxwell, S. Batmanglidj Eckhardt and G. Hite, J. Pharmacol. Exp. Ther. *171*, 62 (1970).
113. R.A. Maxwell, E. Chaplin, S. Batmanglidj Eckhardt, J.R. Soares and G. Hite, J. Pharmacol. Exp. Ther. *173*, 158 (1970).
114. R.A. Maxwell, S.B. Eckhardt, E. Chaplin and J. Burcsu, *In:* Physiology and Pharmacology of Vascular Neuroeffector Systems, p. 349, ed. by J.A. Bevan, R.F. Furchgott, R.A. Maxwell and A.P. Somlyo, S. Karger AG, Basel, 1971.
115. A.S. Horn and S.H. Snyder, J. Pharmacol. Exp. Ther. *180*, 523 (1972).
116. J. Tuomisto, L. Tuomisto and E.E. Smissman, Amer. Med. Exp. Fenn. *51*, 51 (1973).
117. D.D. Miller, J. Fowble and P.N. Patil, J. Med. Chem. *16*, 177 (1973).
118. C.K. Buckner, P.N. Patil, A. Tye and L. Malspeis, J. Pharmacol. Exp. Ther. *166*, 308 (1969).
119. J. Tuomisto, L. Tuomisto and T.L. Pazdernik, J. Med. Chem. *19*, 725 (1976).
120. R.M. Bartholow, L.E. Eiden, J.A. Ruth, G.L. Grunewald, J. Siebert and C.O. Rutledge, J. Pharmacol. Exp. Ther. *202*, 532 (1977).
121. L. Tuomisto, J. Tuomisto and E.E. Smissman, Eur. J. Pharmacol. *25*, 351 (1974).
122. W.L. Nelson, J.E. Wennerstrom, D.C. Dyer and M. Engel, J. Med. Chem. *20*, 880 (1977).
123. W.L. Nelson, M.L. Powell and D.C. Dyer, J. Med. Chem. *22*, 1125 (1979).
124. D.R. Mottram: J. Pharm. Pharmacol. *31*, 767 (1981).
125. R. Weitzell, T. Tanaka and K. Starke, Naunyn-Schmiedeberg's Arch. Pharmacol. *308*, 127 (1979).
126. J.C. McGrath, Biochem. Pharmacol. *31*, 467 (1982).

127. P.S. Portoghese, T.N. Riley and J.W. Miller, J. Med. Chem. *14*, 561 (1971).
128. B. Belleau and P. Cooper, J. Med. Chem. *6*, 579 (1963).
129. B. Belleau and D.J. Triggle, J. Med. Pharm. Chem. *5*, 636 (1962).
130. R.T. Brittain, G.M. Drew and G.P. Levy, Br. J. Pharmacol. *73*, 282P (1981)
131. Y. Nakagawa, N. Shimamoto, M. Nakazawa and S. Imai, Japan. J. Pharmacol. *30*, 743 (1980).
132. E.J. Sybertz, C.S. Sabin, K.K. Pula, G. van der Vliet, J. Glennon, E.H. Gold and T. Baum, J. Pharmacol. Exp. Ther. *218*, 435 (1981)
133. T. Baum, R.W. Watkins, E.J. Sybertz, S. Vemulapalli, K.K. Pula, E. Eynon, S. Nelson, G. van der Vliet, J. Glennon and R.M. Moran, J. Pharmacol. Exp. Ther. *218*, 444 (1981).
134. H.C. Cheng, O.K. Reavis Jr., J.M. Grisar, G.P. Claxton, D.L. Weiner and J.K. Woodward, Life Sci. *27*, 2529 (1980).

Stereoselectivity of Cholinergic and Anticholinergic Agents

R. Dahlbom

Abstract

During the last years much effort has been made to study cholinergic receptors using the stereochemistry of cholinergic and anticholinergic agents as a tool. In this lecture the stereochemistry and stereoselectivity of acetylcholine and its congeners, conformationally restricted analogs of acetylcholine, muscarine, muscarone, dioxolanes and other pentacyclic cholinergic agents, will be discussed as well as some tertiary amines, such as aceclidine and oxotremorine. The antagonists to these types of compounds will also be considered.

The following points will be discussed is some detail.

1. Do all agonists occupy the same binding sites?
2. Do agonists and antagonists occupy the same binding sites?
3. Is the position of a chiral centre in a molecule of importance for the stereoselectivity?

Introduction

More than 75 years ago Cushny observed differences in activity between (–)-hyoscyamine and the corresponding racemate, atropine, and ever since then there has been constant interest in the stereochemistry and stereoselectivity of action of cholinergic agonists and antagonists. Such stereoselectivity may yield information not only on the geometry of the relevant binding sites but also on the relationships between agonist and antagonist sites. A considerable body of information has been accumulated, especially during the last 15 years, and it is impossible to cover it completely. Consequently, I shall confine myself to certain types of cholinergic agonists and antagonists, and the discussion will mainly deal with muscarinic and antimuscarinic agents. Excellent reviews on this subject have been given by Casy[1], Triggle and Triggle[2], and Inch and Brimblecombe[3].

Muscarinic agonists

Methyl acetylcholines. The cholinergic neurotransmitter acetylcholine is achiral and a discussion of its stereoselectivity must therefore be confined to conformational problems. However, since this has already been done in a previous chapter, I shall start with acetylcholine congeners having methyl groups in the choline moiety.

The cholinomimetic activities of the optical isomers of α- and β-methyl-substituted acetylcholine are shown in Table 1. Only one of the four monosubstituted isomers, namely S-(+)-acetyl-β-methylcholine, is equipotent with acetylcholine with respect to muscarinic activity. Apparently the methyl group does not increase the activity, but in the S-β-isomer the interaction with the muscarinic receptor is not hindered, which seems to be the case in the other three isomers. The β-methyl derivatives have little nicotinic activity, but the α-isomers are still potent nicotinic agents. (±)-*Erythro*-acetyl-α,β-dimethylcholine exhibits 14% of the muscarinic activity of acetylcholine, whereas the (±)-*threo*-isomer is practically inert.

Table 1. *Cholinomimetic activities of methyl derivatives of acetylcholine*[4,5]

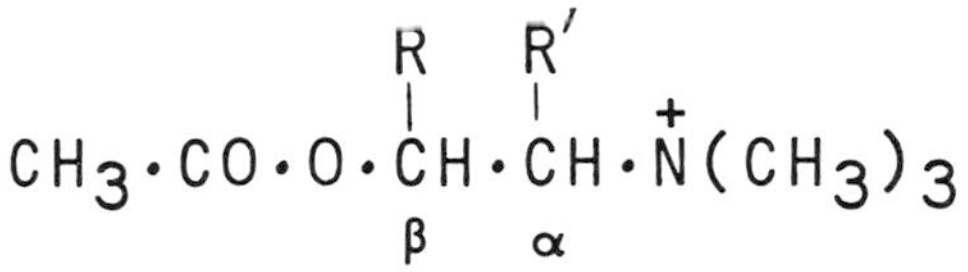

$$CH_3\cdot CO\cdot O\cdot \underset{\beta}{\overset{R}{CH}}\cdot \underset{\alpha}{\overset{R'}{CH}}\cdot \overset{+}{N}(CH_3)_3$$

Configuration	R	R′	EPMR (ACh = 1)		Relative potency
			G.p. ileum	Frog rectus	G.p. ileum
RS	CH_3	H	1.6	180	
S	CH_3	H	1		
R	CH_3	H	240		
RS	H	CH_3	49	2	
S	H	CH_3	232		
R	H	CH_3	28		
(±)-*Erythro*	CH_3	CH_3			14
(±)-*Threo*	CH_3	CH_3			0.036
Acetylcholine	H	H	1		100

Rigid analogues of acetylcholine. Many attempts have been made to build the structural elements of acetylcholine into more or less rigid rings or ring systems in order to 'freeze' the ammonium and ester functions in a given conformation. The trouble with this approach is that the molecule thus obtained must necessarily be bigger than acetylcholine itself, and we have just seen how the addition of a single methyl group can adversely affect the activity. Some interesting rigid compounds have nevertheless been prepared by Cannon *et al.*[6,7]. In order to test the hypothesis that the muscarinic action of acetylcholine is associated with a *trans* conformation and the nicotinic action with a *cis* conformation, the *cis* and *trans* isomers of 2-acetoxycyclopropyltrimethylammonium bromide (**1**,**2**) were prepared. These com-

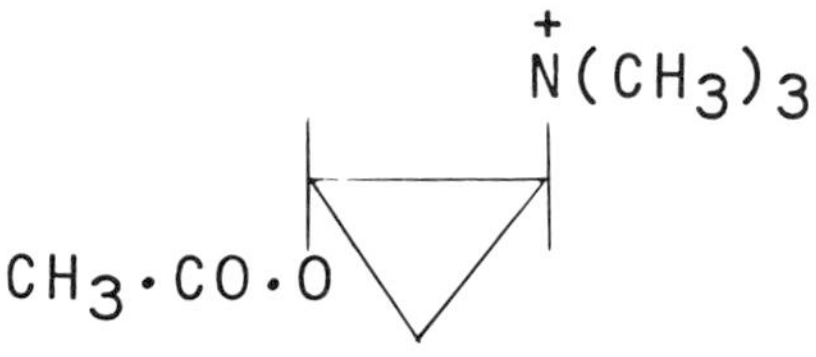

$CH_3\cdot CO\cdot O$ — cyclopropane — $\overset{+}{N}(CH_3)_3$

1

$CH_3\cdot CO\cdot O$ — cyclopropane — $\overset{+}{N}(CH_3)_3$

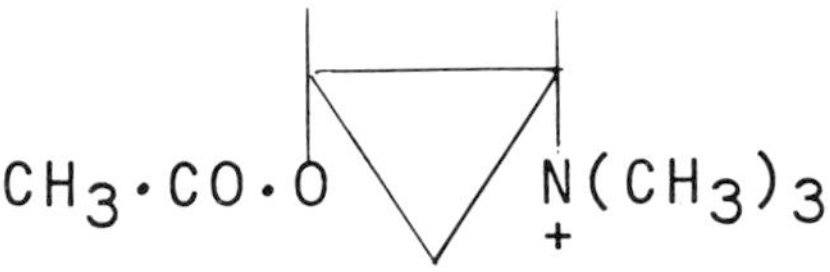

2

pounds contain only one carbon atom more than acetylcholine. It was found that the (+)-isomer of *1* with the absolute configuration 1S, 2S had about the same muscarinic activity as acetylcholine but was a very weak nicotinic agent, whereas the (-)-enantiomer of *1* and the racemate of *2* were virtually inactive.

Muscarine and related compounds. The alkaloid muscarine may be regarded as a cyclic analogue of acetylcholine. Muscarine has three asymmetric carbon atoms and consequently eight optical isomers, all of which have been prepared and investigated[8]. It exhibits a high degree of stereospecificity. Only one isomer, the naturally occurring (+)-muscarine with the configuration 2S, 3R, 5S shows considerable muscarinic activity and it is essentially devoid of nicotinic activity. It would seem that this would greatly facilitate the representation of the topography of the muscarinic receptor, but the situation is apparently not so simple. Oxidation of muscarine destroys the chiral center at the carbon atom in the 3-position and affords muscarone. This operation brings about several important consequences. There is a diminution in the stereoselectivity, as all the isomers of muscarone are more or less active. There is also an inversion of stereoselectivity, the most active isomer of muscarone having the same configuration as the inactive (-)-muscarine isomer (2R, 5R). Moreover, the muscarones exhibit strong nicotinic activity, the most potent in this respect being the racemic form of *allo*muscarone, which has the 2-methyl and the 5-methylammonium groups in the *trans* positions. The cholinomimetic activities of the muscarines and muscarones are shown in Table 2.

Table 2. *Cholinomimetic activities of muscarines and muscarones*[8,9]

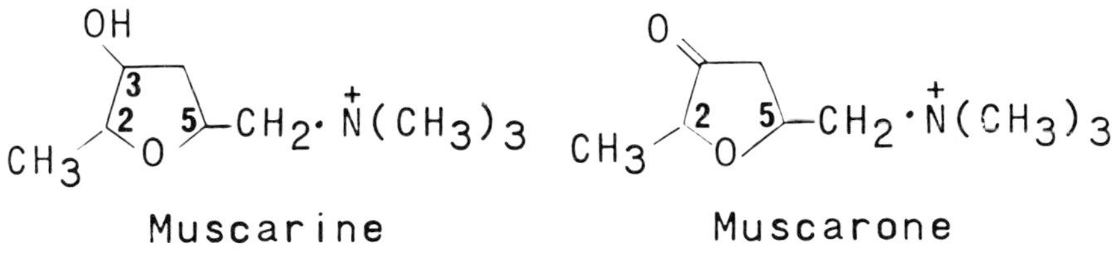

OH, 3, 2, 5, O, CH_3, $-CH_2 \cdot \overset{+}{N}(CH_3)_3$

Muscarine

O, 2, 5, O, CH_3, $-CH_2 \cdot \overset{+}{N}(CH_3)_3$

Muscarone

Compound	Configuration			EPMR (ACh = 1)		
	2	3	5	Cat b.p.	Rabbit ileum	Frog rectus
(+)-Muscarine	S	R	S	0.32	0.33	>50
(-)-Muscarine	R	S	R	350	130	>50
(+)-Muscarone	S		S	0.25	0.15	2.0
(-)-Muscarone	R		R	0.10	0.06	0.5
(±)-Allo-muscarone				0.25	0.28	0.2

Introduction of a trigonal center in the tetrahydrofuran ring of muscarine apparently leads to lower stereoselectivity. This is also the case if the carbon atom in the 5-position is made trigonal by introducing a double-bond. Thus the racemates of both 4,5-dehydromuscarine (*3*) and 4,5-dehydro*epi*muscarine (*4*) are very important muscarinic agents[8], despite the steric relationship of the *epi* compound (*4*) to the inactive *epi*muscarine.

Dioxolanes. A series of compounds closely related to the muscarines are the dioxolanes. The high muscarinic activity of this class of compounds was first discovered in the forties by Fourneau *et al.*, who apparently worked with a mixture of iso-

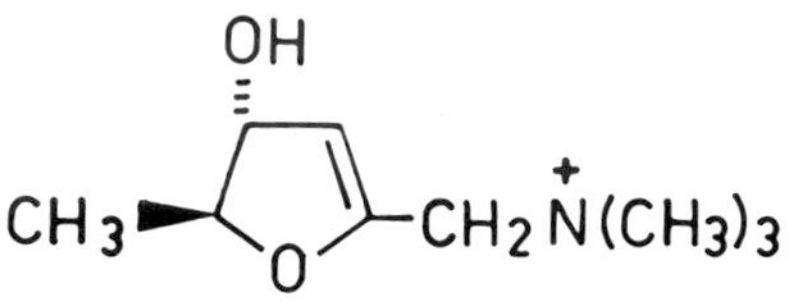

OH

CH_3 — O — $CH_2\overset{+}{N}(CH_3)_3$

3

OH

CH_3 — O — $CH_2\overset{+}{N}(CH_3)_3$

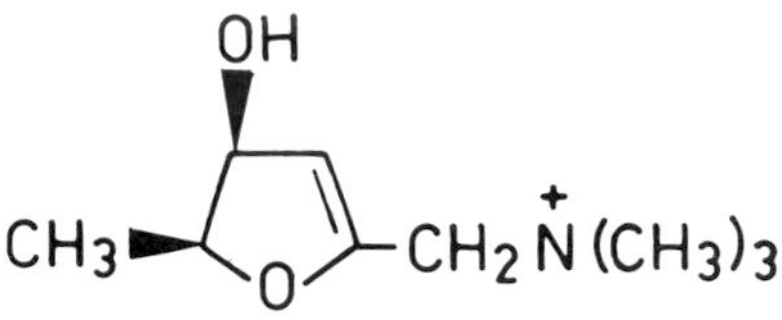

4

mers. The four optical isomers have later been prepared in their pure forms and their stereochemistry has been elucidated by Belleau *et al.*[10–12].

The dioxolanes are very potent specific muscarinic agents, their nicotinic activity being negligible. It is of interest to note that the most active isomer has the substituents in the same relative positions, *cis*, as (+)-muscarine. It is also evident that the configuration of the C-4 site is of greater importance than that of C-2. The muscarinic activities of the four optical isomers of 2-methyl-4-trimethylammoniomethyl-dioxolanes are shown in Table 3.

Table 3. *Muscarinic activities of 2-methyl-4-trimethylammoniomethyldioxolanes*[12]

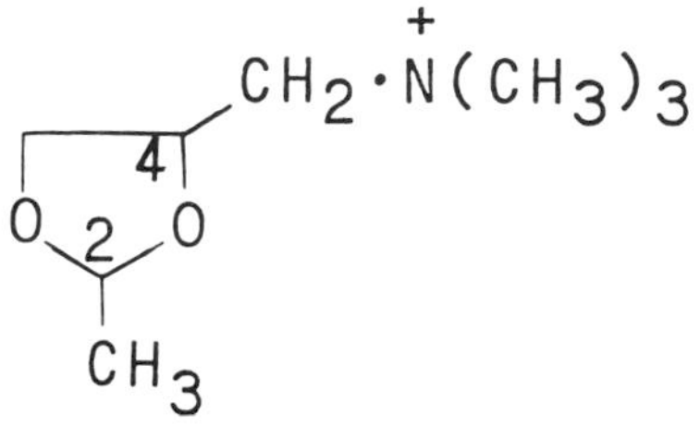

$CH_2 \cdot \overset{+}{N}(CH_3)_3$

4

O 2 O

CH_3

Configuration 2	Configuration 4	Relative potency G.p. ileum
S	R	500
R	S	8.5
S	S	25
R	R	50
Acetylcholine		100

Desether analogues of muscarines, muscarones, and dioxolanes. In order to elucidate the importance of the ethereal oxygen in muscarine and muscarone, this oxygen atom has been changed for a methylene group[13–16]. Surprisingly, these carbocyclic analogues, and especially the muscarone compound, exhibit considerable muscarinic as well as nicotinic activity (Table 4). The muscarone analogue was obtained as a mixture of *cis* and *trans* isomers which could not be separated. All four racemates of deoxamuscarine were prepared, and it is interesting to note that the only racemate which is fairly active has the substituents in the same position as (+)-muscarine. Unfortunately, it has not been possible as yet to resolve this racemate[16].

The same isosteric substitution was made in the 1- and 3-positions of *cis*-1-methyl-4-trimethylammoniomethyldioxolane. Both racemates are active (Table 5) and *7* is a very potent nicotinic agent.

Table 4. *Cholinomimetic activities of desethermuscarones and -muscarines*[16]

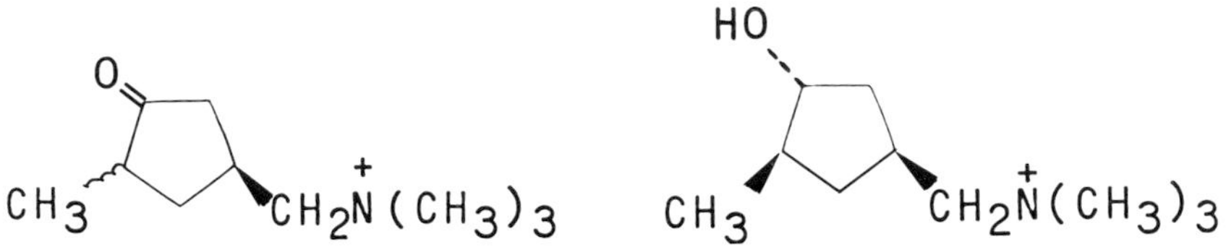

Compound	EPMR (ACh = 1)	
	G.p. ileum	Frog rectus
5 Mixture of (±)-*cis* and *trans* isomers	0.5	2.8
6 (±)	10	75

Table 5. *Cholinomimetic activities of desetherdioxolanes*[16]

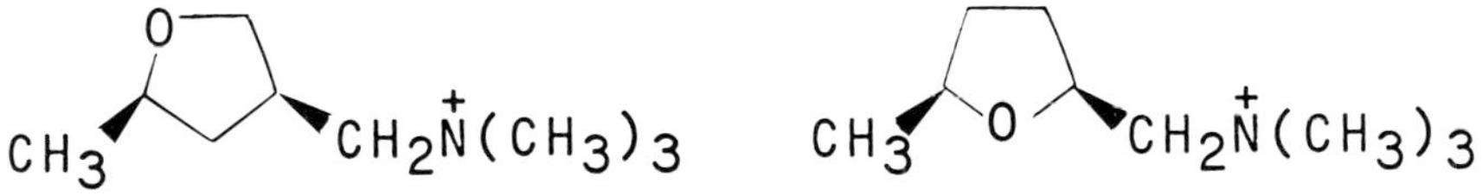

Compound	EPMR (ACh = 1)	
	G.p. ileum	Frog rectus
7 (±)-*cis*	22.8	0.5
8 (±)-*cis*	17.5	38.8

Mode of binding of ligands to the muscarinic receptor. The data presented above show that the stereoselectivity of action and stereochemistry of S-acetyl-β-methylcholine, 2S,3R,5S-(+)-muscarine, and 2S,4R-2-methyl-4-trimethylammoniomethyldioxolane are quite similar. This suggests that these agents bind to the muscarinic receptor in a similar way as the onium group, the methyl and the two oxygens being able to occupy equivalent positions (Fig. 1)[2].

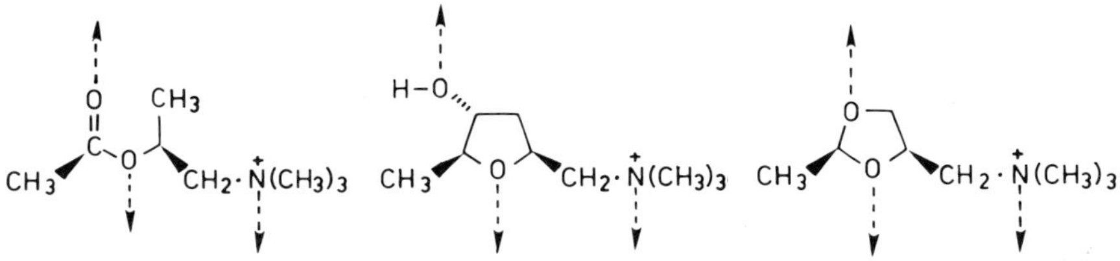

Figure 1. *Interactions of muscarinic drugs with the receptor.*

An alternative mode of binding has been proposed for muscarone and acetyl-α-methylcholine[2,11]. According to this view there is a common binding area which can accomodate the opium and the methyl groups, surrounded by ancillary binding points (Fig. 2). Agents with low stereoselectivity such as muscarone may bind in either way. It is to be regretted that the enantiomers of the desether analogues of muscarine, muscarone, and dioxolanes are not available, as they might give valuable information about the role played by the ethereal oxygen in the mode of binding.

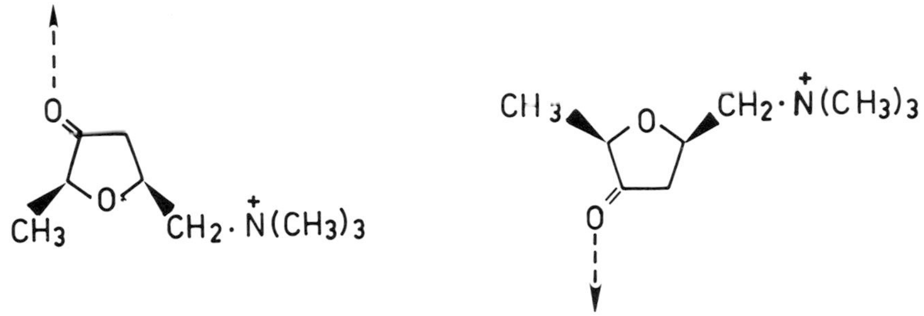

Figure 2. *Alternative modes of interaction with the receptor.*

Antimuscarinic agents

Atropine. Atropine, the racemic form of the alkaloid (-)-hyoscyamine, is the tropyl ester of tropan-3α-ol and is regarded as the archetype for an antimuscarinic agent. From the stereochemical point of view, the most important features are the configuration at the asymmetric carbon atom in the acyloxy moiety and the position of this moiety, α or β, in the tropane ring system. The data in Table 6 show that the most active isomer, (-)-hyoscyamine, has the S-configuration in the tropyl residue, which is attached to the tropane in the α-position.

Table 6. *Antimuscarinic activity of tropyl tropates*[17,18]

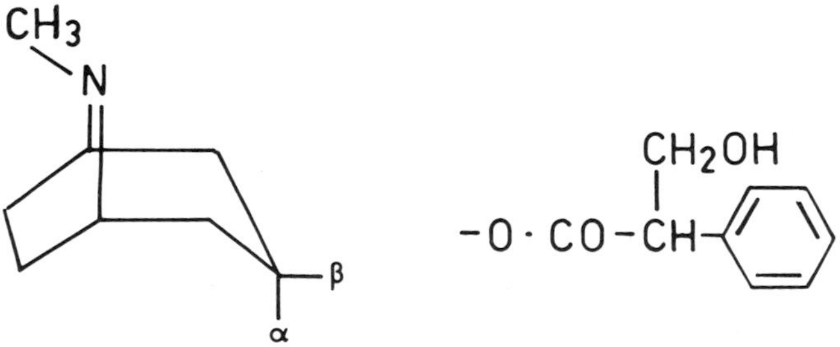

Configuration		G.p. ileum	
		pA_2	log K
α	S		9.4
α	R		6.9
α (Atropine)	RS	9.79	9.0
β	RS	6.55	

Esters of α- and β-methyl-substituted acetylcholine. A number of antimuscarinic drugs have been derived from muscarinic agents by introducing bulky substituents into the agonist. For example, benziloylcholine is a potent antimuscarinic drug and its hexahydro derivative is still more active. In fact, this structure of the acyl residue seems to be optimal for antimuscarinic activity. It is evident that the configuration of the benzylic center in the acid residue is of great importance, the R-enantiomers being the most active (Table 7).

Table 7. *Antimuscarinic activities of esters of choline and α- and β*-methylcholine[19,20]

$$C_6H_5-\underset{R}{\overset{OH}{C}}-CO\cdot O\cdot\underset{\beta}{\overset{R'}{CH}}\cdot\underset{\alpha}{\overset{R''}{CH}}-\overset{+}{N}(CH_3)_3$$

Configuration			R	R′	R″	Rat jejunum pA_2	G.p. ileum log K
Benzylic centre	β	α					
	R		C_6H_5	CH_3	H	8.1	
	S					8.0	
R			C_6H_{11}	H	H	10.4	9.66
S						8.4	7.38
R	R		C_6H_{11}	CH_3	H	8.9	
R	S					8.3	
S	R		C_6H_{11}	CH_3	H	6.9	
S	S					6.7	
R	R		C_6H_{11}	H	CH_3		10.08
R	S						10.04

The esters of α- and β-methylcholine contain a second chriral center and it turns out that the configuration of this is of very little importance, whereas the configuration at the benzylic center is critical to stereoselectivity. We may recall the fact that the parent agonists, α- and β-methyl-substituted acetylcholine, show high stereoselectivity. The affinity of the agonists to the receptor seems to be of no significance with respect to the antagonistic properties. S-Acetyl-β-methylcholine, which is the only really potent agonist in the series, gives rise to the weakest antagonist, and the most potent antagonists are dervatives of the weak agonists R- and S-acetyl-α-methylcholine. The conclusion is that agonists and antagonists do not occupy the same binding sites, or if they do, there must be an ancillary binding area which can strongly bind the acyl moiety in a highly stereoselective manner[21].

2-Substituted dioxolanes. Introduction of bulky groups into the 2-methyl substituent in 2-methyl-4-trimethylammoniomethyl-1,3-dioxolane yields potent muscarinic antagonists. A most admirable piece of work was reported by Brimblecombe and Inch *et al.* who prepared all the eight optical isomers of 2-[(1-cyclohexyl-1-hydroxy-

1-phenyl)-methyl]-4-trimethylammoniomethyl-1,3-dioxolane using stereospecific synthetic methods which yielded the optically pure isomers[22]. The antimuscarinic activities of these compounds are shown in Table 8. It is evident that the configuration at the benzylic center is of the greatest importance, having to be R. It is also of interest to note that the configuration at C-2 is more important than that at C-4, in contrast to the situation in the agonists. Two of the isomers were not tested, probably because of the inactivity of the corresponding teriary amines.

Table 8. *Antimuscarinic activities of 2-[(1-cyclohexyl-1-hydroxyl-1-phenyl)-methyl]-substituted dioxolanes*[22]

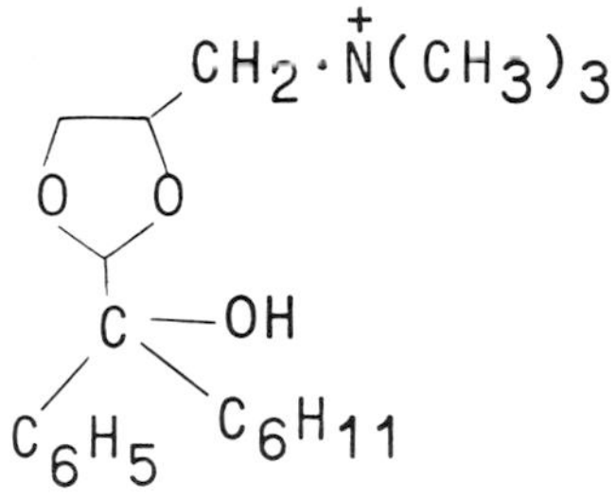

$CH_2 \cdot \overset{+}{N}(CH_3)_3$; O O ; C—OH ; C_6H_5 ; C_6H_{11}

Configuration			
Benzylic centre	2	4	G.p. ileum log K
R	S	R	11.09
R	S	S	9.37
R	R	R	7.60
R	R	S	7.28
S	R	R	6.78
S	R	S	6.55

The most important antimuscarinic drugs contain an asymmetric benzylic carbon atom. It is evident that the groups attached to this carbon atom have the same relative positions in space in the most active enantiomers. Fig. 3 shows the structures of the most active enantiomers of some different types of antimuscarinic agents, and it may be noted that the amino residue, the aromatic ring, and the hydroxyl group occupy the same relative positions. The configuration may be designated R or S, depending on the priority rules used in the nomenclature system.

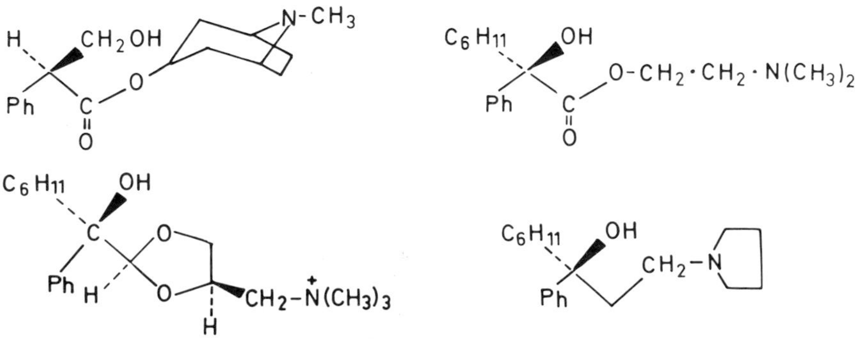

H ; CH_2OH ; $N-CH_3$; O ; Ph ; C ; O
C_6H_{11} ; OH ; $O-CH_2 \cdot CH_2 \cdot N(CH_3)_2$; Ph ; C ; O
C_6H_{11} ; OH ; C ; O ; Ph ; H ; O ; $CH_2-\overset{+}{N}(CH_3)_3$; H
C_6H_{11} ; OH ; CH_2-N ; Ph

Figure 3. *Configuration at the benzylic center in potent antimuscarinic drugs.*

Tertiary muscarinic agonists and their antagonists

The muscarinic agonists discussed so far have all been quaternary ammonium compounds and their tertiary amine analogues possess little or no activity. However, a few tertiary amines are known which are potent muscarinic agents. The alkaloids arecoline and pilocarpine have been known for a long time, and two synthetic compounds, aceclidine and oxotremorine, have in later years been reported to have strong muscarinic properties. These four tertiary amines are considerably more potent than their N-methyl quaternary salts. The reason for this is not quite clear, but it has been suggested that the fact that the nitrogen in these compounds is part of a ring system in a molecule with restricted conformational flexibility may be a contributing factor[23,24].

Quinuclidines. Aceclidine, 3-acetoxyquinuclidine, is a chiral tertiary amine with strong muscarinic activity[25]. The S-enantiomer is about ten times more active than the R-enantiomer[26]; the N-methyl quaternary ammonium salts are much less active. Furthermore, the stereoselectivity is reversed, the R-isomers of the ammonium salts being the most potent[26] (Table 9). The benzilic acid esters of the enantiomers of 3-quinuclidinol, as well as their N-methyl quaternary salts, are potent antimuscarinic agents. However, in this type of compounds the configuration of the amino alcohol is critical, the R-enantiomers being considerably more active than the S-enantiomers[27].

Table 9. *Muscarinic activities of 3-acetoxyquinuclidines*[29]

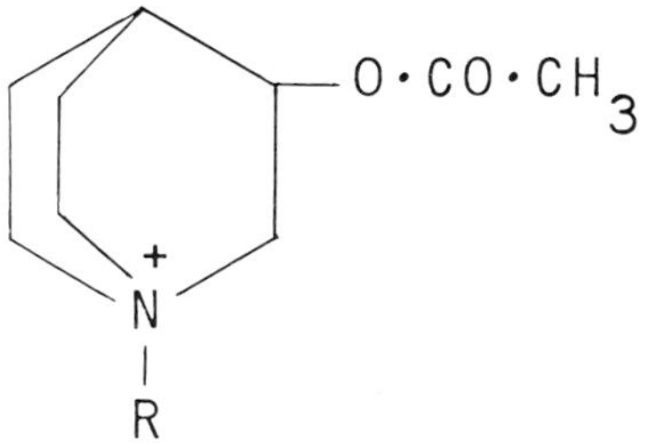

	R	EPMR G.p. ileum
S-(+)-Aceclidine	H	5
R-(–)-Aceclidine	H	71
S-(+)	CH_3	>10000
R-(–)	CH_3	1280
Oxotremorine		1

Inch *et al.* have prepared an ester having two chiral centers, 3R-quinuclidinyl R-hexahydrobenzilate[28]. This compound should be the most active one in this series, and it is in fact the most potent antimuscarinic agent known (Table 10). It would be of great interest to have the remaining three optical isomers investigated.

Oxotremorine. Oxotremorine, N-(4-pyrrolidino-2-butynyl)-2-pyrrolidone (*9*), is a specific muscarinic agent equal in potency to acetylcholine[30]. Being a tertiary amine, it readily penetrates into the central nervous system and produces in experimental animals symptoms resembling those of the parkinsonian patients, including tremors. The structural requirements for muscarinic activity are very specific and even slight changes in structure lead to loss of activity or change of the type of activ-

Table 10. *Antimuscarinic activities of benzilic and hexadrobenzilic acid esters of 3-quinuclidinol*[27,28]

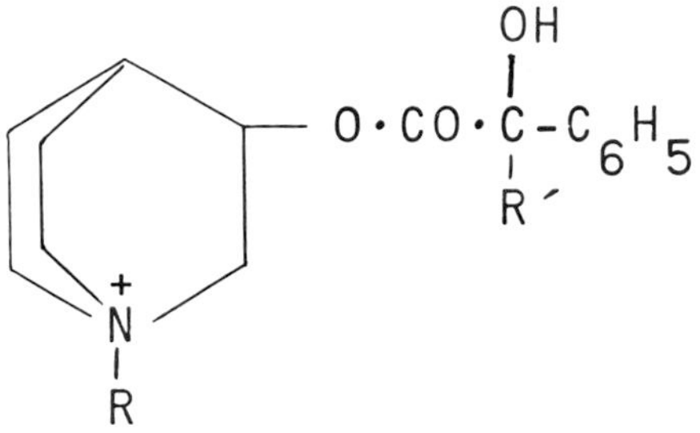

$$\overset{+}{N}(R)\text{-quinuclidin-3-yl}-O\cdot CO\cdot C(OH)(R')-C_6H_5$$

Configuration		R	R′	G.p. ileum	
Benzylic center	Amino alcohol			pA_2	log K
	S	H	C_6H_5	8.21	
	S	CH_3	C_6H_5	8.75	
	R	H	C_6H_5	>10.0	
	R	CH_3	C_6H_5	9.73	
R	R	H	C_6H_{11}		11.67
R	R	CH_3	C_6H_{11}		11.68

ity from agonistic to antagonistic. However, two amino congeners have apprecia-ble muscarinic activity, namely the diethylamino (***10***) and the azetidino (***11***) analogues. The last mentioned compound is in fact the most potent tertiary muscarinic agent known to date (Table 11).

Table 11. *Muscarinic activities of oxotremorine and analogues*[31]

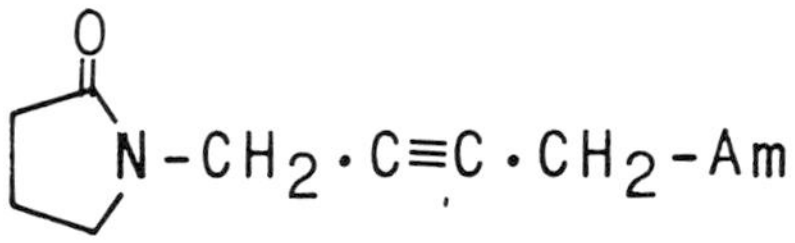

$$\text{2-oxopyrrolidin-1-yl}-CH_2\cdot C\equiv C\cdot CH_2-Am$$

Compound	Am	EPMR G.p. ileum
9	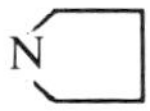 N (pyrrolidino)	1
	 $\overset{+}{N}$(pyrrolidinium)–CH_3	10.9
10	$N(CH_3)_2$	5.0
	$\overset{+}{N}(CH_3)_3$	0.71
11	 N (azetidino)	0.49

A great number of oxotremorine analogues have been prepared in our laboratories[32], and appreciable antagonistic activity to oxotremorine has been observed in several compounds. The most active antagonists contains a methyl group at the C-1 position in a butyne chain. The molecule is now chiral and we prepared the enantiomers in order to study their stereoselectivity. We found that the R-enantiomers were very active, whereas the S-enantiomers were practically inactive[33]. We also prepared analogous compounds substituted with an ethyl or a propyl, and also with both a methyl *and* and ethyl group, in the 1-position of the butynyl chain. Table 12 shows that the R-isomers throughout are more active than the S-isomers. The enantiomeric potency ratio increases with increasing activity of the R-enantiomer (Pfeiffer's rule)[35]. It is interesting to note that higher activity is displayed by analogues with the smaller substituents, the most active being compound ***12***, whose structure differs from that of oxotremorine by only one additional methyl group. This is quite remarkable, as it is completely unique, at least among muscarinic agents, that such a small change in structure should be able to reverse the type of activity and turn a potent agonist into a potent antagonist. We found it also of interest to investigate compounds with a chiral center in the amine part of the molecule, and so we introduced a methyl group in the 2- and 3-positions of the pyrrolidine ring of oxotremorine. In both cases, potent oxotremorine antagonists were obtained. Only the enantiomers of the 2-substituted derivatives show stereoselectivity, the S-enantiomers being the most active (Table 13). However, the stereoselectivity is inferior to that of the enantiomers with the chiral center at C-1 in the butynyl chain.

Table 12. *Tremorolytic activities of N-(1-alkyl-4-pyrrolidino-2-butynyl)-2-pyrrolidones*[33,34]

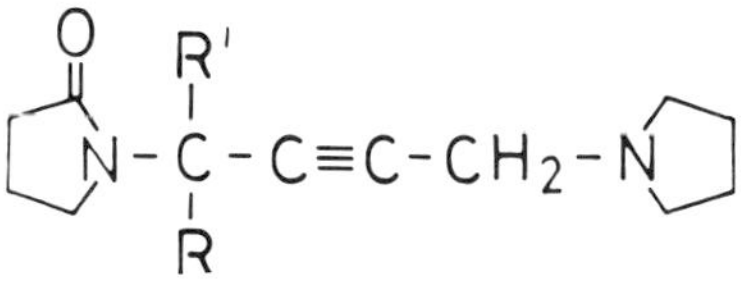

O R'
N-C-C≡C-CH_2-N
R

Compound	R	R′	Tremorolytic dose[a] (μmol/kg)	R/S Potency ratio
R-***12***	H	CH_3	0.26	
S-***12***			Inactive	
R-***13***	H	C_2H_5	0.52	52
S-***13***			27	
R-***14***	H	C_3H_7	3.5	15
S-***14***			51	
R-***15***	CH_3	C_2H_5	28	2
S-***15***			56	

[a] Dose of the test compound required to double the dose of oxotremorine inducing a predetermined tremor intensity in 50% of the mice.

We then studied oxotremorine analogues having both these chiral centers. Eight optical isomers of this type were prepared[37]. The test results of the tremorolytic activity are shown in Table 14. As expected, the isomer having methyl groups at both chiral centers and the 'right' configuration at both centers is the most active, and it is in fact the most potent oxotremorine antagonist currently known.

Table 13. *Tremorolytic activities of N-[4-(methylpyrrolidino)-2-butynyl]-2-pyrrolidones*[36]

Compounds	R	R′	Tremorolytic dose (μmol/kg)	S/R Potency ratio
R-***16***	CH_3	H	56	22
S-***16***			2.6	
R-***17***	H	CH_3	6.5	1.1
S-***17***			5.8	

Table 14. *Tremorolytic activities of N[1-alkyl-4-(2-methylpyrrolidino)-2-butynyl]-2-pyrrolidones*[37]

Compound[a]	R	Tremorolytic dose (μmol/kg)	Enantiomeric dose potency ratio
R,S-***18***	CH_3	0.10	200
S,R-***18***		20	
R,R-***18***	CH_3	0.50	24
S,S-***18***		12	
R,S-***19***	C_3H_7	1.5	16
S,R-***19***		24	
R,R-***19***	C_3H_7	4.2	2.1
S,S-***19***		8.9	

[a] The first configurational symbol refers to the configuration of the chiral center in the butynyl chain and the second symbol to the configuration of the chiral center in the pyrrolidine ring.

In agreement with Pfeiffer's rule, the potency ratios of the enantiomeric pairs increase with increasing activity of the more active isomer of each pair. Epimeric pairs with the opposite configuration in the butyne chain but the same configuration in the pyrrolidine ring display great differences in activity, as shown in Fig. 4. On the other hand, epimeric pairs which have the same configuration in the butyne chain but opposite configuration in the pyrrolidine ring exhibit a small and almost constant potency ratio (Fig. 4). The conclusion which can be drawn from these data is that only the chiral center at C-1 is critical to stereoselectivity, since inversion leads to large differences in potency. The chiral center in the pyrrolidino ring is not critical, since inversion causes only small changes in the epimeric potency ratio. It should be added that the stereochemical requirements and stereoselectivity for antimuscarinic activity in the central nervous system found using this type of test (antagonism to tremors produced by oxotremorine) are the same as in the peripheral nervous system (mydriasis, ileum)[37,38].

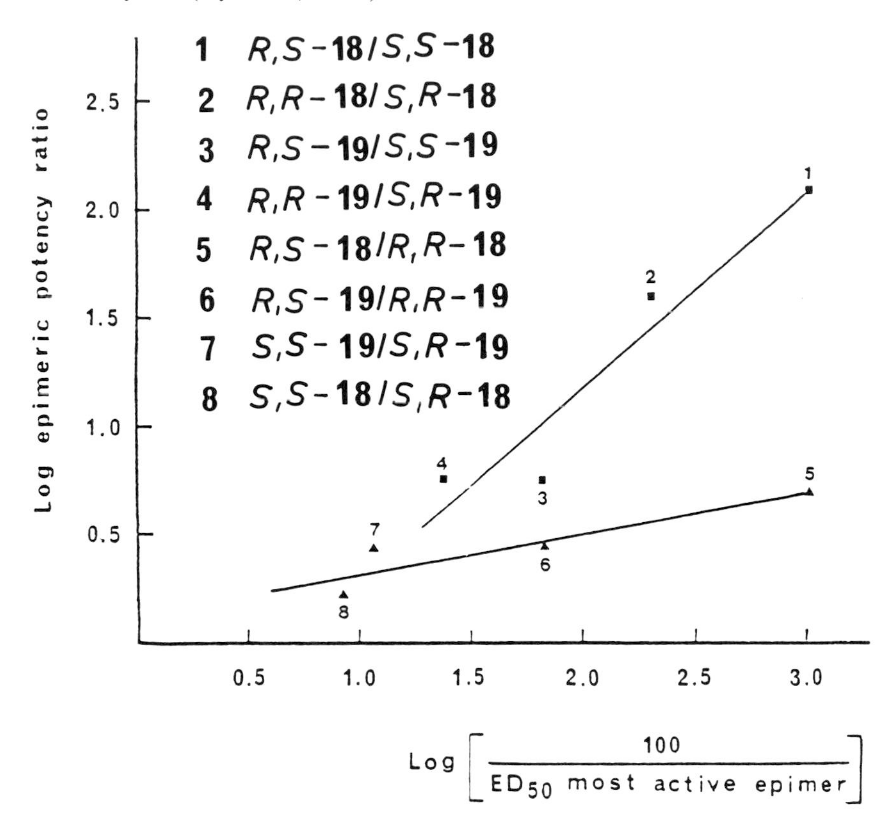

Figure 4. *Tremorolytic activity. The logarithm for the epimeric potency ratio is plotted against the logarithm for the relative activity (100/ED_{50}) of the most active epimer.* ■—— *Compounds epimeric in the butynyl chain;* ▲—— *Compounds epimeric in the pyrrolidine ring.*

As we had in this manner obtained valuable information about the stereoselectivity of antagonists derived from oxotremorine, we found it highly desirable to have access to a chiral agonist in the oxotremorine series for comparative purposes. It was not feasable to obtain such agonists by introducing substitutents into the oxotremorine molecule to make it chiral, for reasons already explained. However, already in 1966 Bebbington *et al.* prepared and tested an oxotremorine congener, namely N-4-pyrrolidino-2-butynyl)-N-methylacetanilide (***20***), and found it to be a potent tremorogenic agent about half as active as oxotremorine[39]. Considerable activity was also shown by its dimethylamino analogue (***21***). We became interested in this type of compounds and prepared a number of analogues for structure-activity studies[40]. We found indeed what we were looking for – introduction of a methyl group at C-1 in the butynyl chain of ***20*** and ***21*** yielded two compounds with very interesting properties[41]. The pyrrolidino compound was an antagonist to oxotremorine in the CNS, but on guinea pig ileum it was a partial agonist. The dimethylamino analogue was a pure agonist both centrally and peripherally. The enantiomers of the two compounds and their N-methyl quaternary salts were prepared, and

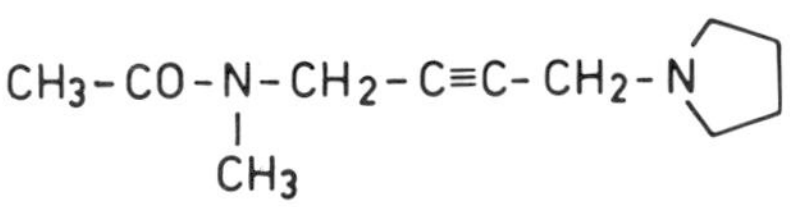

$CH_3-CO-N(CH_3)-CH_2-C{\equiv}C-CH_2-N$ (pyrrolidine ring)

20

$CH_3-CO-N(CH_3)-CH_2-C{\equiv}C-CH_2-N(CH_3)_2$

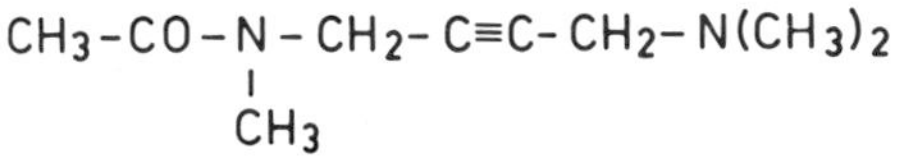

21

it was found that the R-enantiomers were more potent than the S-isomers both *in vivo* and *in vitro* regardless of whether the compounds were agonists, partial agonists, or antagonists (Table 15). Thus the stereochemical requirements for muscarinic and antimuscarinic activity seem to be similar. This finding, together with the close structural similarity between agonists and antagonists make it seem probable that in the oxotremorine series agonists and antagonists bind to a common receptor site. However, when antimuscarinic agents of the classical type are used in test systems with oxotremorine agonists, the antagonists show the usual stereoselectivity and the configuration of the benzylic center is critical[20,28].

Table 15. *Pharmacological activities of N-(4-amino-1-methyl-2-butynyl)-N-methylacetanilides*[41]

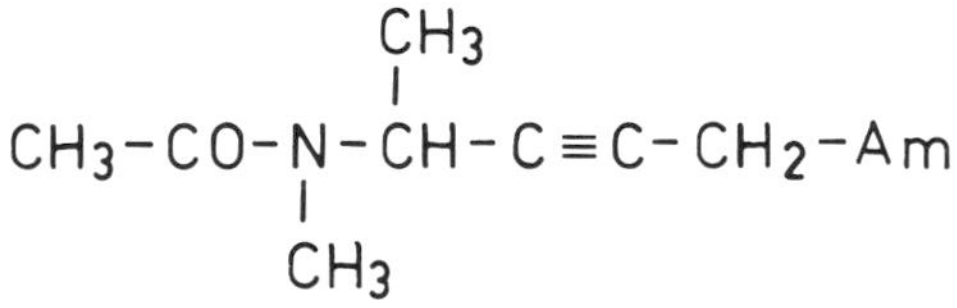

$$CH_3-CO-N(CH_3)-CH(CH_3)-C{\equiv}C-CH_2-Am$$

Compound	Am	Guinea pig ileum pD_2	Tremorogenic dose[b] (μmol/kg)	Tremorolytic dose (μmol/kg)
R-*22*	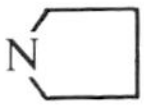 N (pyrrolidine ring)	6.63[a]		0.35
S-*22*		5.51[a]		10
R-*23*	$N(CH_3)_2$	6.50	12	
S-*23*		5.29	inactive	
R-*24*	$\overset{+}{N}(CH_3)_3$	6.80		
S-*24*		5.28		
Oxotremorine		7.50	0.5	
Carbachol		7.15		

[a] Partial agonist.
[b] Dose required to induce a predetermined tremor intensity in mice.

Conclusions

Most of the antimuscarinic agents known are considered to be competitive antagonists of acetylcholine and similar agonists, and it would be reasonable to assume that the agonist and the antagonist share at least one common binding site. However, already in 1926 Clark[42] suggested that atropine and acetylcholine do not interact at the same site. It might be possible that atropine acts at a regulatory site separated from the agonist binding site but linked to it in such a way that the binding of atropine induces a change in the affinity of the agonist binding site. Several facts are known which support this interpretation and the most important pieces of evidence have been summarized by Casy[1]:

1. An increase in the size of the cationic head of acetylcholine reduces cholinergic activity; similar changes in an antagonist have little effect on potency. With a few exceptions a quaternary nitrogen is mandatory for potent agonist activity but antagonists may have either a tertiary or a quaternary nitrogen.
2. The great influence of α- or β-methyl substitution on the agonist properties of acetylcholine is in sharp contrast to the minor effects of such changes on the properties of choline esters with antagonistic action; in such esters the stereochemistry of the acyl moiety is of paramount importance.
3. In 1,3-dioxolanes with muscarinic properties, the configuration of the carbon in the 4-position has a greater influence upon activity than that in the 2-position. However, the stereochemistry at C-4 is of little importance in dioxolanes which block acetylcholine; if they carry a benzylic substituent at C-2, potency is chiefly governed by the configuration of the benzylic carbon.

Thus it seems highly improbable that agonists and antagonists bind at the same receptor sites.

However, derivatives of 3-quinuclidinol and oxotremorine may form exceptions. In the quinuclidine derivatives the configuration in the amino alcohol part of the molecule is highly important both for agonistic and antagonistic potency, though the stereochemical relationships are not quite clear. In the oxotremorine series the steric requirements for agonistic and antagonistic activity appear to be similar. In these cases the agonists and antagonists may interact with a common receptor site. Be that as it may, much work remains to be done before the question as to whether muscarinic and antimuscarinic agents act on the same or on different receptors can be answered.

References

1. A.F. Casy, Prog. Med. Chem., *11,* 1 (1975).
2. D.J. Triggle and C.R. Triggle, Chemical Pharmacology of the Synapse, Academic Press, New York, p. 291 (1976).
3. T.D. Inch and R.W. Brimblecombe, Internat. Rev. Neurobiol., *14,* 67 (1974).
4. A.H. Beckett, N.J. Harper and J.W. Clitherow, J. Pharm. Pharmacol., *15,* 362 (1963).
5. E.E. Smissman, W.L. Nelson, J.B. LaPidus and J.L. Day, J. Med. Chem., *9,* 458 (1966).
6. P.D. Armstrong and J.G. Cannon, J. Med. Chem., *13,* 1037 (1970).
7. C.Y. Chiou, J.P. Long, J.P. Cannon and P.D. Armstrong, J. Pharmacol. Exp. Ter., *166,* 243 (1969).
8. P.G. Waser, Pharmacol. Rev., *13,* 465 (1961).
9. L. Gyermek and K.R. Unna, Proc. Soc. Exp. Biol. N.Y., *98,* 882 (1958).
10. D.J. Triggle and B. Belleau, Can. J. Chem., *40,* 1201 (1962).
11. B. Belleau and J. Puranen, J. Med. Chem., *6,* 325 (1963).
12. B. Belleau and J.L. Lavoie, Can. J. Biochem., *46,* 1397 (1968).
13. K.G.R. Sundelin, R.A. Wiley, R.S. Givens and D.R. Rademacher, J. Med. Chem., *16,* 235 (1973).
14. R.S. Givens and D.R. Rademacher, J. Med. Chem., *17,* 457 (1974).
15. F. Gualtieri, M. Giannella, C. Melchiorre and M. Pigini, J. Med. Chem., *17,* 455 (1974).
16. F. Gualtieri, P. Angeli, M. Giannella, C. Melchiorre and M. Pigini, in Recent Advances in Receptor Chemistry (F. Gualtieri, M. Giannella and C. Melchiorre, eds.), Elsevier/

North-Holland Biomedical Press, Amsterdam, p. 267 (1979).
17. R.B. Barlow, F.M. Franks and J.D.M. Pearson, J. Med. Chem., *16,* 439 (1973).
18. R.J. Hunt and J.B. Robinson, J. Pharm. Pharmacol., *24,* 324 (1972).
19. B.W.J. Ellenbroek, R.J.F. Nivard, J.M. van Rossum and E.J. Ariëns, J. Pharm. Pharmacol., *17,* 393 (1965).
20. R.W. Brimblecombe, D.M. Green, T.D. Inch and P.B.J. Thompson, J. Pharm. Pharmacol., *23,* 745 (1971).
21. E.J. Ariëns and A.M. Simonis, Ann. N.Y. Acad. Sci., *144,* 842 (1967).
22. R.W. Brimblecombe, T.D. Inch, J. Wetherell and N. Williams, J. Pharm. Pharmacol., *23,* 649 (1971).
23. I. Hanin, D.J. Jenden and A.K. Cho, Mol. Pharmacol., *2,* 352 (1966).
24. A.K. Cho, D.J. Jenden and S.I. Lamb, J. Med. Chem., *15,* 391 (1972).
25. M.D. Mashkovsky and L.N. Yakhontov, Prog. Drug. Res., *13,*, 293 (1969).
26. R.B. Barlow and A.F. Casy, Mol. Pharmacol., *11,* 690 (1975).
27. G. Lambrecht, Eur. J. Med. Chem., *14,* 111 (1979).
28. T.D. Inch, D.M. Green and P.B.J. Thompson, J. Pharm. Pharmacol., *25,* 359 (1973).
29. B. Ringdahl, F.J. Ehler and D.J. Jenden, Mol. Pharmacol., *21,* 594 (1982).
30. A.K. Cho, W.L. Haslett and D.J. Jenden, J. Pharmacol. Exp. Ther., *138,* 249 (1962).
31. B. Resul, B. Ringdahl, R. Dahlbom and D.J. Jenden, Eur. J. Pharmacol., in press.
32. R. Dahlbom, in Cholinergic Mechanisms: Phylogenetic Aspects, Central and Peripheral Synapses, and Clinical Significance (G. Pepeu and H. Ladinsky, eds.), Plenum Press, New York, p. 621 (1981).
33. R. Dahlbom, Å. Lindquist, S. Lindgren, U. Svensson, B. Ringdahl and M.R. Blair, Jr., Experientia, *30,* 1165 (1974).
34. B. Ringdahl and R. Dahlbom, Acta Pharm, Suec., *16,* 13 (1979).
35. E.J. Ariëns, Adv. Drug. Res., *3,* 235 (1966).
36. B. Ringdahl and R. Dahlbom, Experientia, *34,* 1334 (1978).
37. B. Ringdahl, B. Resul and R. Dahlbom, J. Pharm. Pharmacol., *31,* 837 (1979).
38. B. Ringdahl and D.J. Jenden, Mol. Pharmacol., in press.
39. A. Bebbington, R.W. Brimblecombe and D. Shakeshaft, Brit. J. Pharmacol., *26,* 56 (1966).
40. B. Resul, R. Dahlbom, B. Ringdahl and D.J. Jenden, Eur. J. Med. Chem. *17,* 317 (1982).
41. R. Dahlbom, D.J. Jenden, B. Resul and B. Ringdahl, Brit. J. Pharmacol., *76,* 299 (1982).
42. A.J. Clark, J. Physiol. *61,* 547 (1926).

Stereoselectivity in Dopaminergic and Antidopaminergic Agents

M.G. Bogaert and W.A. Buylaert

Abstract

Many dopamine agonists and antagonists show optical stereoselectivity, and stereoisomers have been used extensively for the study of the dopamine receptor. Dopamine itself has no center of asymmetry but is a flexible molecule, and rigid or semi-rigid analogues containing the dopamine segment have been studied, in order to find the receptor preferred dopamine conformation.

On the basis of studies with stereoisomers, with semi-rigid or rigid analogues, and with other agonists and antagonists, different classifications of the dopamine receptors, based on ligand binding studies, biochemical assays or pharmacological tests, have been proposed, as well for the central nervous system as for the periphery. Important discrepancies between these classifications are apparent. More systematic studies with the different isomers and with agonists and antagonists in general are needed.

Introduction

It is now recognized that dopamine receptors are present in behavioral, gastrointestinal, cardiovascular and endocrine systems. On the basis of studies with agonists and antagonists it has been proposed that several subtypes of dopamine receptors exist and a number of classifications of these subtypes have appeared in the literature.

Kebabian and Calne[1] have introduced a biochemical classification, with D_1-dopamine receptors linked to an adenylate cyclase and D_2-dopamine receptors not linked to an adenylate cyclase. Seeman[2] proposed a classification of ligand binding sites in the central nervous system in D_1, D_2, D_3 and D_4, with the D_2 binding site ('sensitive to micromolar concentrations of dopamine but nanomolar concentrations of neuroleptics') corresponding to the dopamine receptor. Cools and Van Rossum[3] suggest a classification of 'excitatory' and 'inhibitory' dopamine receptors, which originates from the properties of dopamine receptors in the snail Helix Aspersa. Goldberg[4] suggests for the classification of peripheral dopamine receptors in the cardiovascular system, DA_1 (postsynaptic) and DA_2 (presynaptic).

It is not our purpose to review all the data that led to these classifications and the discussions thereabout[5,6]. We intend only to give a brief review of some data on stereoselectivity of dopamine agonists and antagonists which have contributed to the characterization of the dopamine receptors.

Stereoisomers of agonists and antagonists

In contrast to dopamine itself which has no center of asymmetry, a number of dopamine agonists present *optical isomerism*. Apomorphine e.g. exists as a levorotatory and a dextrorotatory isomer. In 1973, Saari et al.[7] showed that the dextrorotatory enantiomer 6aS (+)-apomorphine is devoid of agonistic (or antagonistic) properties at central dopamine receptors, while 6aR (-)-apomorphine is a much used dopamine agonist. Optical stereoselectivity has also been found for a number of dopamine antagonists, e.g. the neuroleptic butaclamol. The stereoselectivity of a neuroleptic such as butaclamol is used routinely in ligand binding studies to distinguish between specific and non-specific binding. The optical stereoselectivity of dopamine agonists and antagonists has however only a limited role in helping to classify different subtypes of dopamine receptors; indeed, with few exceptions (see below), optical stereoselectivity applies to all dopamine receptors, whatever their localization or their presumed function. Likewise optical stereoselectivity cannot be used to decide whether an effect is mediated by dopamine receptors, as a similar degree of stereoselectivity can exist for other receptors[8].

For some agonists and antagonists, there is *geometrical isomerism* and the cis-and trans-forms are very different in regard to their potency for influencing dopamine receptors.

There has been much interest for the receptor preferred conformation of dopamine (Fig. 1). Dopamine exists in two gauche forms and one staggered or trans form; it is generally accepted that the trans form is preferred by the dopamine receptor, and this in its coplanar or trans β conformation[9,10].

For the trans β conformation, α- and β-rotamers are possible, depending on the orientation of the catechol ring in relation to the ethylamino side chain. With a flexible molecule such as dopamine, one cannot be sure that the conformation one observes in a solution or in a crystal corresponds to the receptor preferred conformation, and therefore conformationally rigid or semi-rigid molecules which contain the dopamine segment have been synthetized.

These molecules have been extensively tested to approach the problem whether the α-rotameric or the β-rotameric trans β conformation of dopamine is preferred by the receptor. In Fig. 2 some of the dopamine analogues containing the α- and β-rotameric dopamine conformation are shown.

Stereoselectivity at central dopamine receptors

Seeman[2] recently reviewed extensively the data on brain dopamine receptors and proposed, on the basis of binding characteristics of dopamine agonists and antagonists, the existence of 4 binding sites (D_1, D_2, D_3, D_4). By comparing these results with the results of biochemical and pharmacological tests, Seeman concludes that only the D_2 binding site fulfills all the criteria for a receptor, and he uses the word 'receptor' only in connection with the D_2 site. Fig. 3 is taken from Seeman's work based on the structure-binding data for dopamine agonists acting on that receptor. Some of these data are derived from work with stereo-isomers and with rigid or semi-rigid analogues. The requirement that the nitrogen atom of the dopamine agonist be positioned at 0.6 Å from the plane of the ring is in accordance with the fact e.g. that levorotatory 6aR (-)-apomorphine is active, while its 6aS (+)-dextrorotatory isomer is not[7]. The distance between the hydroxyl group and the nitrogen group should be 7.3 Å or less; this could explain some results obtained with the aminotetralines, e.g. why for 2-amino-6,7-dihydroxytetraline, the (+)-isomer with

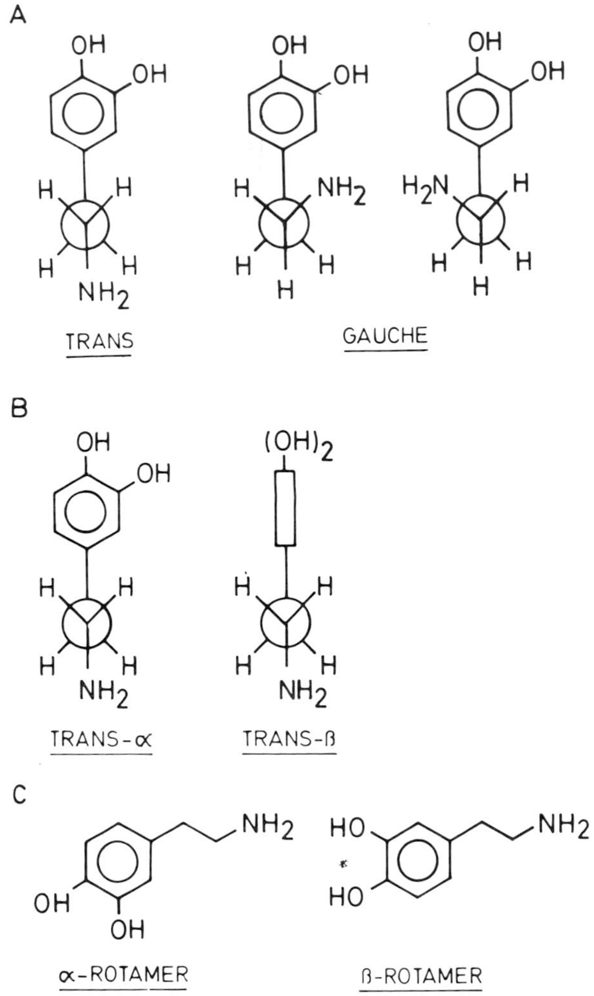

Figure 1. *Some dopamine conformations.*
A : Gauche and trans (or staggered) conformations
B : α- and β-forms of the trans conformation
C : α- and β-rotameric forms of the trans β conformation.

a distance of 7.3 Å, is much more active than the (-)-isomer for which this distance is 7.8 Å (±)-2-amino-6,7-dihydroxytetraline, containing the dopamine segment in a β-rotameric conformation, is several times more potent on Seeman's D_2-receptor than its 5,6 analogue which contains the α-rotameric dopamine segment.

While Seeman's[2] classification is based on ligand binding of agonists and antagonists, there have been, as already said, other classifications, based e.g. on biochemical criteria such as adenylate cyclase activity or on a number of pharmacological tests. The results obtained with stereoisomers in these different experimental systems are not always concordant, and the discussion about the activity of the α-rotameric versus the β-rotameric aminotetralines is a good illustration of these discrepancies. Indeed with ligand binding studies[2] and with adenylate cyclase measurement[11], β-rotameric analogues were found to be more active; in a number

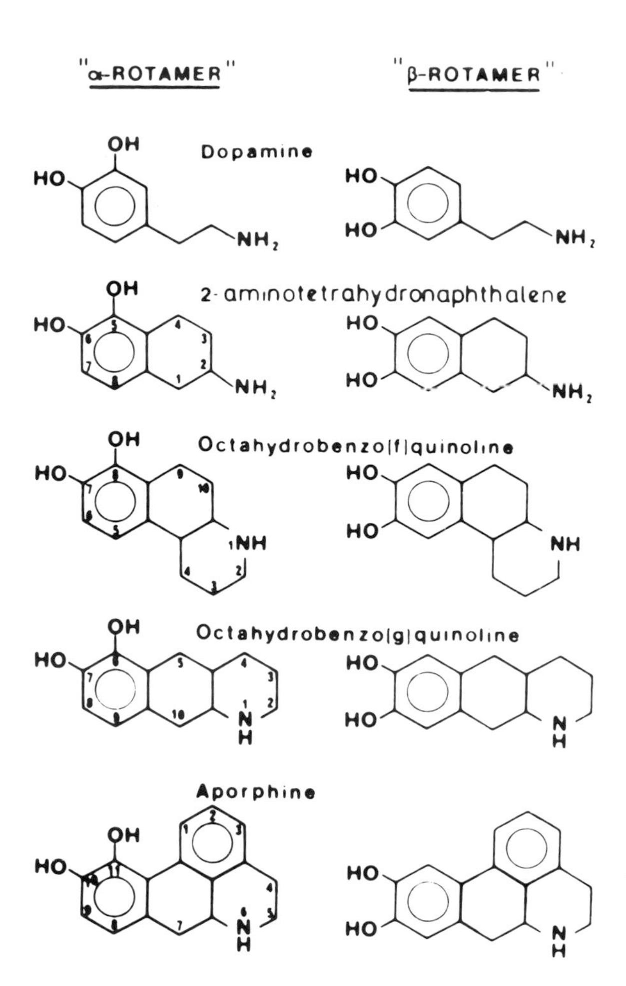

Figure 2. *α- And β-rotameric conformation of dopamine and structural analogues of different series with the dopamine segment in the α- and β-rotameric conformation (from B. Costall, S.K. Lim, R.J. Naylor, and J.G. Cannon, J. Pharm. Pharmacol. 34, 246, 1982, with permission of the authors and of the journal).*

of pharmacological tests in the central nervous system, the reverse was true[10]. Some authors, Seeman e.g.[2,5], argue that the best comparison of potencies of dopamine congeners on brain dopamine receptors can be done when the true concentration at the receptor level is known and this is impossible when pharmacological tests are used. Differences in physicochemical properties of various dopamine agonists will indeed affect passage through the blood brain barrier; although this can be avoided by injecting the drug directly in the central nervous system, even then liposolubility will influence distribution within a brain area. The differences in breakdown of e.g. α- and β-rotameric analogues in a given series, could also affect the results obtained in pharmacological tests, although this probably does not explain the discrepancies cited earlier[10]. It is however clear that the affinity found in a ligand binding study, depends on the composition of the incubation buffer and other experimental factors, which makes extrapolation to the in vivo situation dangerous.

The question whether or not there is more than one type of central dopamine receptor is at this moment therefore certainly not answered.

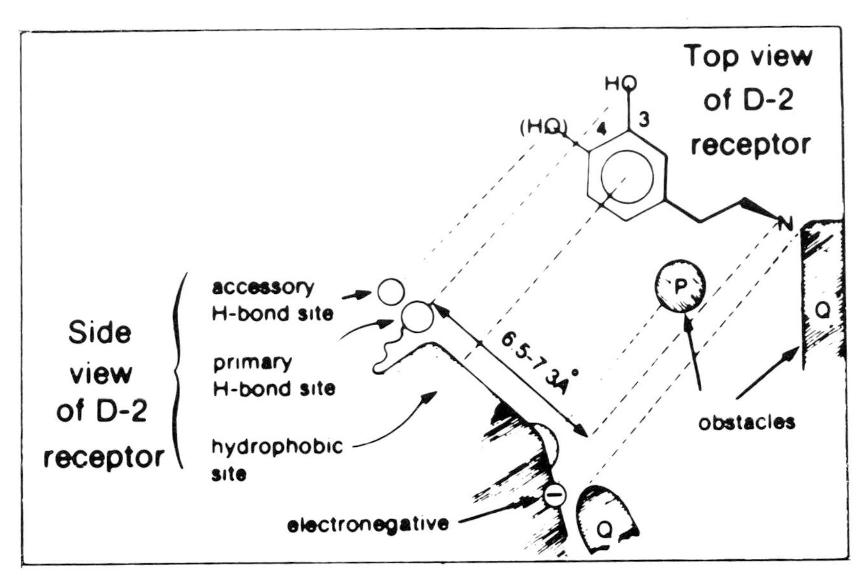

Figure 3. *Interpretation of the structure-activity data for dopamine agonists acting on the D_2 dopamine receptor (from Ph. Seeman, Pharmacol. Rev. 32, 229, 1980; with permission of the author and of the journal).*

Stereoselectivity at peripheral dopamine receptors

There is now good evidence for the existence of ganglionic, presynaptic and postsynaptic dopamine receptors e.g. in the cardiovascular system[4,12]. From work with agonists and with antagonists it is apparent that presynaptic and postsynaptic receptors are different[4]. Some of the evidence comes from work on the stereochemical requirements for both agonist and antagonists.

It has been suggested that the α- and β-rotameric semi-rigid analogues of dopamine differ markedly in their potency for pre- and for postsynaptic peripheral dopamine receptors, with both 5,6 and 6,7-aminotetralines active on the presynaptic (DA_2) receptor and only the 6,7 (β-rotameric) analogue active on the postsynaptic (DA_2) receptor[13]. However, the same authors have now shown that dipropyl substitution in the N-position of the 5,6 analogue, confers a potent DA_1-agonist activity[14], and other authors did not find a marked difference in potency for DA_1 and DA_2 between 5,6 and 6,7 unsubstituted aminotetralines[15].

For dopamine antagonists, as a rule stereoselectivity applies similarly for presynaptic and for postsynaptic peripheral receptors, and corresponds to the stereoselectivity observed in the central nervous system (Fig. 4). However there are exceptions. S(–)-sulpiride is more active than R(+)-sulpiride on presynaptic dopamine receptors, as shown in the femoral artery[13], the ear artery[16] and the perfused cat spleen[17]. In contrast, when tested on the renal artery postsynaptic dopamine receptor, R(+)-sulpiride was found to be more active[16].

The difference between pre- and postsynaptic peripheral dopamine receptors is also apparent from work done with piribedil, with ergot alkaloids and with different neuroleptics[4].

Central versus peripheral dopamine receptors

The above mentioned results with sulpiride, and the work with other antagonists and agonists have led to the suggestion that the presynaptic peripheral dopamine receptor would be more closely related to the dopamine receptor or receptors in the

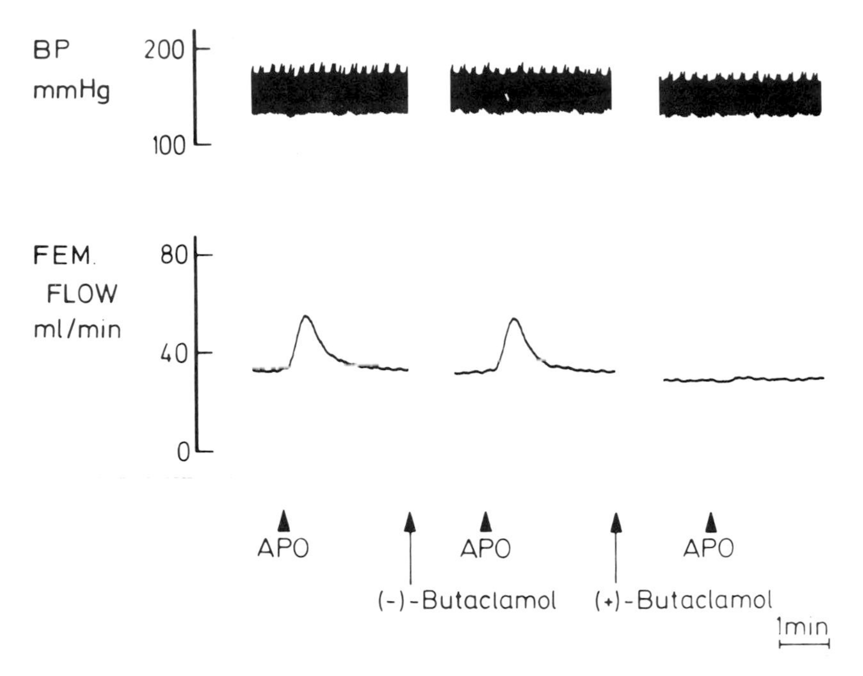

Figure 4. *Influence of apomorphine (0.25 × 10^{-8} mole) given into the femoral artery of the dog, on blood pressure (BP) and femoral artery flow (FEM. FL.). Antagonism of the effect of apomorphine by (+)-butaclamol (14 × 10^{-8} mole), but not by (–)-butaclamol (14 × 10^{-8} mole).*

central nervous system than the postsynaptic peripheral vascular receptor[4]. However, such a conclusion is certainly premature and a number of discrepancies still remain. It should be mentioned that for the peripheral dopamine receptor the number of substances tested systematically is small as compared to the central nervous system. Moreover, in vitro data on peripheral dopamine receptors are scarce, and ligand binding studies are only just starting[18].

Conclusions

Much work has been done on the stereochemical requirements for dopamine agonists and antagonists in different systems, as well with optical and with geometrical isomers, as with rigid or semi-rigid analogues containing the dopamine segment in one conformation or another. Discrepancies are apparent between results obtained with ligand binding studies, with biochemical assays and with pharmacological tests, and there is no agreement on the different classifications of dopamine receptors which have been proposed. One can hope that a more systematic study of agonists and antagonists will lead to a better knowledge of the dopamine receptors, also in the peripheral systems. This could result in the development of compounds with a selective effect, allowing e.g. for agonists inhibition of prolactin secretion or improvement of parkinsonism without eliciting vomiting and nausea or for antagonists, a neuroleptic effect without extrapyramidal side-effects.

References

1. J.W. Kebabian and D.B. Calne, Nature (London) *277*, 93 (1979).
2. Ph. Seeman, Pharmacol. Rev. *32*, 229 (1980).
3. A.R. Cools and J.M. van Rossum, Life Sci. *27*, 1237 (1980).

4. L.I. Goldberg and J.D. Kohli, In 'Apomorphine and Other Dopaminomimetics, Vol. 1: Basic Pharmacology', p. 273 (G.L. Gessa and G.U. Corsini, eds.). Raven Press, New York, 1981.
5. Ph. Seeman, Biochem. Pharmacol. *31*, 2563 (1982).
6. I. Creese and D.R. Sibley, Biochem. Pharmacol. *31*, 2568 (1982).
7. W.S. Saari, S.W. King, and V.J. Lotti, J. Medicin. Chem. *16*, 171 (1973).
8. S.J. Enna, J.P. Bennett, D.R. Burt, I. Creese, and S.H. Snyder, Nature (London) *263*, 338 (1976).
9. A.S. Horn and J.R. Rodgers, J. Pharm. Pharmacol. *32,* 521 (1980).
10. B. Costall, S.K. Lim, R.J. Naylor, and J.G. Cannon, J. Pharm. Pharmacol. *34*, 246 (1982).
11. G.N. Woodruff, K.J. Watling, C.D. Andrews, J.A. Poat, and J.D. McDermed, J. Pharm. Pharmacol. *29*, 422 (1977).
12. W.A. Buylaert, J.L. Willems, and M.G. Bogaert, In 'Vasodilatation', p. 125 (P.M. Vanhoutte and I. Leusen, eds.). Raven Press, New York, 1981.
13. L.I. Goldberg and J.D. Kohli, Comm. Psychopharmacol. *3*, 447 (1979).
14. J.D. Kohli, L.I. Goldberg, and J.D. McDermed, Eur. J. Pharmacol. *81,* 293 (1982).
15. R.A. Brown, R.C. Brown, S.E. O'Connor, and A.M. Solca, Brit. J. Pharmacol. *67*,421P (1979).
16. R.A. Brown and S.E. O'Connor, Brit. J. Pharmacol. *73*, 189P (1981).
17. M.L. Dubocovich and S.Z. Langer, J. Pharmacol. Exp. Ther. *212,* 144 (1980).
18. O.-E. Brodde, Life Sci. *31*, 289 (1982).

Stereoselectivity in Peptides

A. Witter

Abstract

Amino acids are the basic structural units of peptides. The α-carbon in amino acids is asymmetric and two stereoisomers (L and D) of practically all amino acids exist. Virtually all of the amino acids present in peptides from animal sources are the L-(natural) isomers; only in plants and bacteria some D-amino acids occur. Peptides that are to contain D-amino acids must therefore be prepared synthetically, using the desired D-amino acid(s) as starting compounds in the synthesis. The effect of D-amino acid substitutions on the properties of peptides are two-fold: 1. conformational changes; D-amino acid substitutions influence the projection of side chains, that play a role in tertiary structure (conformation) and receptor interaction (biological activity profile). 2. peptide bonds susceptible to enzymatic attack are often rendered more resistent by D-amino acid substitutions. Such substitutions can prolong the duration of action, render oral administration effective and direct metabolic pathways towards intermediate peptides with individual biological activity profiles. Ultimate goals are increased stability and specificity. D-amino acid substitutions have been applied for various peptides. Allretro-D-analogs, built entirely of D-amino acids bonded in reversed sequence were found biologically inactive, despite semi-equivalent orientation of the side chains. This suggests a role for the backbone of peptide bonds in receptor interactions as well. D-amino acid substitutions in LH-RH agonists and antagobists increase potency and reveal opposite effects related to dosage because of interactions at both the pituitary and gonadal level. D-amino acid substituted enkephalins possess suitable potencies but lack a much desired increased specificity. In ACTH neuropeptides D-amino acid substitutions increase potency and duration of action and shift behavorial activity profiles. In vasopressin D-amino acid substitutions can increase specificity by preferentially decreasing the vasopressor/antidiuretic activity ratio.

Asymmetry of amino acids

Amino acids, the basic structural units of peptides, consist of an amino group (-NH_2), a carboxyl roup (-COOH), a hydrogen atom (-H) and a distinctive side chain (–R), all bonded to a carbon atom, called by the α-carbon, The α-carbon of all approximately 20 different amino acids is asymmetric, except glycine (R = H). Therefore two stereo-isomers of each amino acid exist and these two mirror-image forms are colled the L-isomer and the D-isomer. It is the tetrahedral array of four different groups around the α-carbon atom that makes amino acids optically active. The absolute configurations of amino acids are referred to that of serine ($R = CH_2OH$). The optical rotation of L-serine is laevo (left): (–) or (l), but many other L-amino acids are dextro (right): (+) or (d). The sign for optical rotation is usually omitted, because the degree and sometimes even the sign of rotation change with the composition (i.c. pH) of the solution. Two amino acids (isoleucine and threonine) possess a second asymmetric carbon atom - in the side chain - called the β-carbon. For these amino acids two possible isomers exist in both the L- and D-series (the allo forms).

Occurrence in nature

All of the amino acids present in peptides from animal sources are the L-isomers. Only recently an opioid peptide from amphibian skin, called dermorphin, containing a D-amino acid residue (D–alanyl–), has been isolated[1]. Some D-amino acid residues occur in plants and bacteria. For example D-alanine and D-glutamic acid occur in bacterial cell walls. Many antibiotics contain D-amino acids, sometimes unusual ones (cf. D-penicillamine in penicillin). The allo forms have not been found in animal tissues but allothreonine is present in phalloidine, a peptide from the mushroom Amanita phalloides.

It is usual to omit the prefix L: all references to amino acids imply the natural or L-isomers, unless otherwise specified.

Nomenclature of peptides

Peptides are composed of chains of amino acids. The amino acids are combined by peptide bonds - CO-NH -, formed by loss of a watermolecule following the joining of an α-amino group of one amino acid to an α-carboxyl group of another amino acid. An amino acid in a peptide chain is called an amino acid residue and is usually abbreviated, either by a three-letter abbreviation (often the first three letters) or (particularly in proteins) by a one-letter symbol. The residue with the free α-amino group (if present) is placed at the left of the sequence (the NH_2-terminal) and numbered 1.

Peptides containing D-amino acid residues

All peptides obtained from animal sources, with presently a single exception[1], contain natural or L-amino acid residues. This implicates that peptides that are to contain D-amino acid residues must be prepared synthetically, using the desired D-amino acid (S) as starting compounds in the synthesis. An alternative route, mainly of theoretical interest, is heat-alkali treatment of natural peptides. The heating of natural peptides in the presence of sodium hydroxide leads to partial racemisation, that is the conversion of the L- to the D-enantiomer, of some amino acid residues in the peptide structure. The melanocytestimulating hormones α- and β-MSH have been extensively investigated in this respect. Although the original observation dates back as far as 1924, even today the phenomenon is not fully understood[2].

Properties of peptides containing D-amino acid residues

The original observations on the effect of racemisation of the MSH's on biological activity demonstrated a prolongation of the biological effect and either a retardation of the response (α-MSH) or an immediately enhanced and potentiated response

(β-MSH). It was suggested that configurational changes in the peptide, as a consequence of chemical changes because of heat-alkali treatment, led to altered affinity for receptor sites and/or changes in rates of degradation.

Also the present state of the art attributes a two-fold effect of D-amino acid substitutions on the (bio)chemical properties of the peptide in question.

1. *Conformational changes*

The structure of peptides can be considered at different levels:

primary structure:	the sequence of amino acids in the peptide chain;
secondary structure:	the formation of the peptide chain in a spiral structure (α-helix) as a consequence of hydrogen bonding between peptide bonds;
tertiary structure:	the 3-dimensional folding of the secondary spiral structure, imposed by intramolecular forces (ionic; hydrophobic; dipole-, hydrogen-, disulfide bondes; etc.) between side chains (R) of the amino acid residues.

In particular the secondary and tertiary structures are relatively unstable because they are maintained by rather weak bonds. Moreover, in oligopeptides – being relatively small molecules – the intramolecular forces are low in number and therefore play a relatively small role. Consequently, oligopeptides appear to be strongly flexible in solution and can assume different conformations with comparable energy. Additionally, conformations in solution are affected by the nature of the microenvironment (pH, interaction of reactive-R groups with solute molecules). Even in the solid state relatively little is known of the three-dimensional structure of peptides. The effect of substitutions of L- by D-amino acid residues on the conformation is even more obscure. Nevertheless, it seems quite propable that such substitutions can strongly influence conformation, for example in the projection of the side chains. Since the side chains play an important role in the tertiary structure as well as in receptor interactions, significant changes at the receptor level and subsequently in the biological activity profiles could result. Although it might be assumed that substitutions in the region essential for biological activity will contribute most to such changes, the outcome of D-amino acid substitutions on conformation and subsequently biological activity still remains academic guess-work.

2. *Changes in metabolic stability*

The enzymes involved in catalyzing the hydrolysis of peptide bonds ultimately yield the free constituing amino acids. The hydrolysis proceeds through the formation of smaller, intermediate oligopeptides. These intermediates may be more or less stable, more or less potent and may have different profiles of biological activity. The situation is similar to that seen in the biotransformation of non-peptide drugs. It seems reasonable to assume that proteolytic enzymes evolved using peptides composed of L-amino acids as their natural substrates. Proteinases generally distinguish between optical isomers: a susceptible peptide bond is often rendered resistant to attack by substitution to one or both amino acid residues adjoining the susceptible peptide bond by the D-isomer. The biological half-life of natural peptides is short, usually in the order of minutes, and increased metabolic stability can increase the half-life considerably. However, peptides are attacked by a multiplicity of proteolytic enzymes and multiple D-amino acid substitutions might be necessary to obtain a significant increase in half-life. Besides increasing overall stability, D-amino acid substitutions can direct the metabolic pathway and lead to the accumulation of intermediate peptides with increased metabolic stability. If such substitutions are carried out in the region essential for biological activity the resulting intermediate peptides could retain biological activity and a prolongation of biological effect would result. An increase in the duration of biological activity could thus result from an increase in stability of the parent peptide and/or a biologically active intermediate peptide. Moreover, the biological activity spectrum of the parent and/or the intermediate peptide could be changed – through conformational changes – as a result of the substitution. Changes in biological activity spectra should preferably increase specificity. It is conceivable that (intermediate) oligopeptides with less structural units available for interaction with various potential receptor sites, could be of special interest in this respect.

Applications of D-amino acid substituted peptides

The main concern of substitutions with D-isomers is the development of more stable, longer-acting peptide derivatives (parent or intermediate) with increased specificity, that could be administered by oral or similar routes.

There are at least ten different biologically active peptides of which the effect of D-amino acid substitutions has been investigated during the last decades. The present state of the art appears at the threshold between experimental development and practical applicability, which might be illustrated by the following examples:

1. *Retro-D-analogs*

An interesting approach has been the synthesis of a so called retro-D-analog of a peptide, a peptide chain built entirely of D-amino acids bonded in the reverse sequence as the L-peptide. In a retro-D-analog the overall shape and topochemical pattern of the side chains are quasi-equivalent to that of the L-peptide (provided proline is absent in the sequence). On the other hand, the retro-D-analog differs from the L-peptide in the reversal of the C=O and H-N elements of the peptide bonds and in the reversal of the $-NH_2$ and – COOH termini. Assuming that the biological activity is primarily related to the side chains, retro-D-analogs with quasi-equivalent shape and patterns of the side chains should exhibit biological activity. Furthermore, the retro-D-analogs should be metabolically stable, retain activity when administered orally and have a prolonged duration of action. These predictions were not substantiated by experimental evidence. Neither retro-D-bradykinin nor retro-D-oxytocin analogs were found biologically active (ref. 3). These results suggest that not only the side chains but also certain elements of the peptide backbone are involved in receptor-occupation and/or receptor-activation, since reversal of the building elements of this backbone results in loss of biological activity.

2. *Luteinizing hormone-releasing hormone (LH-RH)*

This decapeptide of hypothalamic origin has the following primary structure:

$$\text{p-}\overset{1}{\text{Glu}}\text{–}\overset{2}{\text{His}}\text{–}\overset{3}{\text{Trp}}\text{–}\overset{4}{\text{Ser}}\text{–}\overset{5}{\text{Tyr}}\text{–}\overset{6}{\text{Gly}}\text{–}\overset{7}{\text{Leu}}\text{–}\overset{8}{\text{Arg}}\text{–}\overset{9}{\text{Pro}}\text{–}\overset{10}{\text{Gly}}\text{–NH}_2$$

The N- and C-termini of this peptide are not free: the NH_2 terminus is present as an internal amide bond, a pyrrolidone carboxylic acid (p-Glu) residue and the – COOH terminus as the amide – $CONH_2$.

LH-RH exerts a direct action on the pituitary, stimulating the secretion of luteinizing hormone (LH) and follicle stimulating hormone (FSH). These hormones in turn stimulate gonadal functions. Thus LH-RH is indispensable in follicular maturation, ovulation, corpus luteum function and pregnancy and it has been succesfully applied in women suffering from hypothalamic amenorrhea. The therapeutic application of LH-RH is limited by its lack of activity when given orally, its short duration of action and lack of complete specificity. In order to overcome these limitations D-substituted analogs were prepared, primarily aimed at increasing the stability of the most susceptible peptide bonds[4]. LH-RH is only slowly degraded by plasma, but is rapidly inactivated by tissue extracts including liver, kidney, brain and the pituitary. The latter is a target tissue for LH-RH and metabolism in this tissue could be an important factor in regulating the concentration of biologically active LH-RH available for receptor interaciton. Primary cleavage sites of LH-RH appear to the -$\overset{6}{\mathrm{Gly}}$-$\overset{7}{\mathrm{Leu}}$- and the -$\overset{9}{\mathrm{Pro}}$-$\overset{10}{\mathrm{Gly}}$-$NH_2$ bonds. Consequently, LH-RH analogs with modifications at the positions 6, 7, 9 and 10 were investigated and found to be more potent and longer acting than the parent peptide. Essentially, such analogs contain a D-amino acid residue in position-6 (D-Ala-, D-Leu-, D-Trp-, D-Tyr, D-Phe-) replacing the original glycine residue in combination with an alkyl group replacing the glycinamide residue in position-10; modifications at position-7 are less frequent[4]. The availability of radioactively labelled LH-RH analogs and the development of radioligand ('receptor') assays allowed a more detailed comparison between the parent peptide and D-substituted analogs. Such investigations made it likely that increased metabolic stability was only partly responsible for the observed changes in biological activity. Although the analogs were found to be considerably more stable during *in vitro* incubations in tissue extracts, they also showed increased pituitary uptake, longer retention in the pituitary and higher affinity for specific binding sites obtained from hypophyseal tissue. This demonstrates an important role of higher pituitary receptor binding, probably resulting from conformational changes, in the increased biological activity as well.

Such highly potent LH-RH analogs have been applied to the treatment of human infertility. However, upon repeated administration a striking decrease in responsiveness was observed, indicating a difference in action with LH-RH. It appeared that this decreased responsiveness could only partly be explained by desensitization of pituitary receptors. When higher dosages of these potent LH-RH analogs were applied in order to overcome this apparent tolerance, a paradoxical phenomenon was observed. The analogs, initially developped as agonists for the promotion of fertility, elicited anti-fertility effects - that is antagonistic activity - using the higher dosages. It became apparent that these highly potent LH-RH analogs not only exert a direct action at the pituitary level but also at the gonadal level, causing an inhibition of gonadal steroidogenesis, ovulation and gametogenesis, ovum implantation and pregnancy. These effects are opposite to those mediated by LH and FSH at the gonadal levels and thus opposite to the effects of LH-RH at the pituitary level[5]. The gonadal binding sites responsible for this unexpected effect of LH-RH showed characteristics similar to those observed in the pituitary: highly potent analogs exhibited significantly higher affinities than LH-RH itself[6]. The concentration of circulating hypothalamic LH-RH seems too low to make a physiological action on gonadal LH-RH receptors likely. On the other hand, the presence of specific ovarian LH-RH receptors not linked to a physiological function is rather unexpected. Local secretion of LH-RH in control of ovarian function therefore seems likely. The existence of ovarian LH-RH receptors and the inhibitory effects of pharmacological doses of D-substituted LH-RH agonists on reproductive processes has opened new potentials for the use of LH-RH agonists in the prevention or interception of implantation as well as in the regression of mammary tumors[7]. However, the practical use in contraception is complicated by conflicting reports on side effects, that could appear as a consequence of disturbed follicular function.

Another class of potential fertility-control agents comprises LH-RH antagonists, LH-RH analogs that inhibit the release of LH and FSH at the pituitary level[4]. These LH-RH analogs act as competitive LH-RH antagonists, that is they possess reduced intrinsic activities and compete with LH-RH at the pituitary level. Structural modifications resulting in reduced efficiency at the receptor level involve histidine in position-2 and to a lesser extent tryptophane in position-3. LH-RH antagonists include des-His^2-LH-RH and 2-substituted analogs like [Phe^2] – , [Tyr^2] – and [Trp^2] -

LH-RH. As in the case of LH-RH agonists substitution of the original Gly^6-residue by D-amino acids [D-Aka^6] - etc. potentiates the antagonistic properties of the His^2-modified LH-RH analogs[4]. Such analogs show increased metabolic stability. On the other hand, the antagonistic activities of these analogs are closely related to their binding affinities for the LH-RH pituitary receptor sites. This again indicates that the effects of D-amino acid substitutions at the level of receptor interaction are at least as important as those resulting in increased metabolic stability.

LH-RH antagonists have been found to inhibit ovulation and their application as contraceptive, following intranasal administration, has been advocated.

The LH-RH antagonists not only displace LH-RH (agonists) from pituitary, but also from ovarian binding sites. Therefore they should be able to antagonize the effect of pharmacological doses of LH-RH agonists at the gonadal level as well[8]. The complexities arising from opposite effects related to dose have delayed practical application of LH-RH agonists and antagonists.

3. *Enkephalins*[9]

The opioid peptides [Leu^5] enkephalin and [Met^5] enkephalin are pentapeptides of brain origin.

$$\overset{1}{\text{H-Tyr}}\text{-}\overset{2}{\text{Gly}}\text{-}\overset{3}{\text{Gly}}\text{-}\overset{4}{\text{Phe}}\text{-}\overset{5}{\text{Leu}}\text{-OH} \qquad \overset{1}{\text{H-Tyr}}\text{-}\overset{2}{\text{Gly}}\text{-}\overset{3}{\text{Gly}}\text{-}\overset{4}{\text{Phe}}\text{-}\overset{5}{\text{Met}}\text{-OH}$$

These peptides exert morphine like actions and were regarded as potential morphinomimetic drugs. The clinical usefulness of morphine is limited by its large number of side effects, like respiratory depression, neuro-endocrine changes, development of tolerance and physical dependence and its use as an addictive substance. It was hoped that enkephalin analogs could be developed that would act more specifically.

The development of analogs was virtually imperative, because the enkephalins are highly unstable in a biological environment. The effect of amino acid substitutions was mainly tested *in vitro* using biological and opioid binding assays. It was soon recognized that the tyrosine in position-1 is essential for opioid activity. This is not restricted to the whole tyrosine molecule: removal or replacement of the p-hydroxyl group or the amino group are detrimental for agonist activity. Also configurational requirements are very precise: the [D-Tyr^1] enkephalins are inactive. In view of this and the extreme lability of the H-Tyr^1-Gly^2- bond it is evident that hydrolysis of this peptide bond, resulting in release of the N-terminal tyrosine, results in complete inactivation. Efforts to stabilize this labile bond are necessarily restricted to modifications in position-2. Introduction of D-amino acid residues in this position did cause a marked increase in bio-potency. With [Leu^5] enkephalin optimal increase in potency (about 50 times) is observed by replacement of glycine-2 by D-serine; residues of similar size, D-Ala, D-Thr- and D-Met-, potentiate 15-40 times. With [Met^5] enkephalin the effect of D-2-amino acid substitution is less pronounced and in fact only the [D-Ala^2] analog is more potent than the parent peptide. These data indicate that besides stabilization of the labile peptide bond also configurational elements contribute to the observed potentation. The latter is also obvious from increased affinity of D-2-substituted enkephalins in receptor bindings, although this is considerably less pronounced than in the case of LH-RH.

Investigations concerned with D-amino acid substitutions in other positions are limited. Replacement of glycine-3 by D-Ala- or D-Pro- and of phenyl-alanine-4 by D-Ala-, D-Phe- or D-Trp- result in pronounced reduction of biological activity. Although numerous modifications of the C-terminal residue have been carried out, the number of D-amino acid residue substitutions is restricted to D-Leu-, D-Met- and D-Ala-. These D-5- substituted analogs are biologically active, albeit usually less than the parent compound. The results of D-amino acid substitutions in position-1 through 5 demonstrate again their impact on conformation and subsequently on biological activity at the level of the receptor, although in this case their most important contribution appears on peptide bond stability, in particular the H-Tyr^1-Gly^2-bond. This is corroborated by the high potency of analogs in which this -CO-NH- bond is replaced by metabolically more stable non-peptide bonds like -CO-CH_2- or - CH_2-NH-. As before, the fully retro-enantiomer H-D-Met-D-Phe-Gly-Gly-D-Tyr-OH was inactive.

The clinical applicability of the enkephalins is of course of considerable interest. For opioid peptides analgesia seems an obvious choice in this respect. The predictive value of *in vitro* data for *in vivo* analgesic potency, obtained from tail-flick or hot plate model tests, appeared to be low. This was first ascribed to rapid degradations of enkephalins by brain enzymes. D-Ala^2-substituted analogs that are metabolically considerably more stable did approach the potency of morphine following intracerebroventricular administration. However, after peripheral administration such analogs were hardly active and this was attributed to failure of the peptides to penetrate the blood-brain-barrier. Multiple modified enkephalins, invariably including a D-amino acid substitution in position-2 and usually combined with C-terminal modifications, with potencies up to a 1000-fold as compared to morphine using intracerebroventricular administration are now available. These peptides are generally 1-10 times more potent than morphine after i.v. or s.c. administration and some are comparable to morphine even after oral administration. It appears that analgesic potency and metabolic stability must be satisfied for high potency. These are not restricted to enhanced or improved receptor binding or increased specificity for a particular opioid receptor subtype (μ-receptors versus δ-receptors), but might include phenomena as bioavailability, transport (including binding to plasma proteins) and ability to cross the blood-brain-barrier (possibly by facilitating transport mechanisms).

The development of enkephalins with suitable potencies however represents only one side of the clinical applicability medal. The other side concerns specificity. In this respect the more serious side effects of morphine are of principal interest. It is as yet too early to draw definite conclusions, but the outlooks for highly specific opioid peptides are not too favourable. Structural modifications that increase potency do not generally increase specificity of the enkephalin analogs. Physical dependence and respiratory depression correlate approximately with analgesic activity, cardiovascular effects appear variable, neuro-endocrine responses conflicting. The latter could be related to the involvement of different opioid receptor subtypes. Knowledge of the stringent structural and configurational requirements for specific receptor interactions is apparently essential for a rational development of specifically acting enkephalins. This must probably await the availability of pure opioid receptor populations.

4. *ACTH-Neuropeptides*

Peptides derived from the N-terminal part of the ACTH molecule

$$\overset{1}{\text{H-Ser}}\text{-}\overset{2}{\text{Tyr}}\text{-}\overset{3}{\text{Ser}}\text{-}\overset{4}{\text{Met}}\text{-}\overset{5}{\text{Glu}}\text{-}\overset{6}{\text{His}}\text{-}\overset{7}{\text{Phe}}\text{-}\overset{8}{\text{Arg}}\text{-}\overset{9}{\text{Trp}}\text{-}\overset{10}{\text{Gly}}\text{-...}$$

affect avoidance behavior but are virtually devoid of classical endocrine activities. In particular ACTH-(4-10)-heptapeptide, [D-Phe^7] ACTH-(4-10)-heptapeptide and the analog [Met(O_2)^4, D-Lys^8, Phe^7] ACTH-(4-9)-hexapeptide (Org 2766) have been extensively investigated. The ACTH-(4-10)-peptide, the reference peptide, inhibits extinction of conditioned pole-jumping avoidance behavior and facilitates passive avoidance responding.

Both *in vitro* and *in vivo* studies have shown that the peptide bonds $-\overset{6}{\text{His}}-\overset{7}{\text{Phe}}-\overset{8}{\text{Arg}}-\overset{9}{\text{Trp}}-$ are susceptible to protolytic attack. Substitution of the phenylalanine residue in position-7 by its D-isomer could be expected to affect proteolysis and [D-Phe^7] ACTH-(4-10) was indeed found to possess a prolonged duration of action. However, also a shift in the behavioral activity profile was observed: [D-Phe^7] ACTH-(4-10) facilitates both behaviors mentioned. The threefold substituted analog Org 2766 is approximately 1000 × more potent than the ACTH-(4-10)-peptides and its duration of action is similar to that of [D-Phe^7] ACTH-(4-10). The behavioral provile of Org 2766 however is similar to that of ACTH-(4-10). Thus, substitution of -Phe^7- by its D-enantiomer increases duration of action and shifts the behavioral profile, substitution of -Arg^8- by a D-lysine residue (in combination with two additional substitutions not involving D-residues) increases similarly the duration of action, does not shift the behavioral profile but markedly increases potency. The concurrent introduction of D-methionine in position-4 and of D-phenylalanine in position-7 reverses the behavioral activity profile: [D-Met^4, D-Phe^7, Lys^8] ACTH-(4-9) inhibits extinction of active avoidance responding in a similar way as ACTH-(4-10). These observations suggest a loop-(helix) like structure in which methionine-4 comes close to phenylalanine-7. Obviously, intrinsic activity as well as metabolic properties are affected by these substitutions[10].

At increased dose levels an opposite effect of Org 2766 on passive avoidance responding was observed. This shift is exclusive for Org 2766 and passive avoidance responding, since Org 2766 inhibited active extinction as before and neither ACTH-(4-10) nor [D-Phe^7] ACTH-(4-10) showed a shift at higher dosages. Evidence was obtained that the dual effect of Org 2766 on passive avoidance responding is due to two opposite behavioral activities located in the same molecule. The N-terminal Met(O_2)^4 ACTH-(4-7)-tetrapeptide facilitates, the C-terminal [D-Lys^8, Phe^9] ACTH-(7-9)-tripeptide inhibits passive avoidance responding. It might be speculated that the tripeptide, because of the presence of a D-amino acid residue in position-8, is metabolically more stable than the tetrapeptide and accumulates relative to the tetrapeptide. This is substantiated by earlier findings showing that, following *in vitro* incubation of [D-Lys^8] ACTH-(4-9)-hexapeptide, the main metabolite was the C-terminal [D-Lys^8] ACTH-(7-9)-tripeptide[11]. It is conceivable that this relative accumulation of the tripeptide becomes more pronounced at highter doses and is responsible for the observed shift in passive avoidance responding[12]. In that case D-amino acid substitution not only concerns overall metabolic stability but can also direct metabolism preferentially towards the metabolically most stable intermediate peptide. If such a preferred metabolite possesses a different biological activity profile as compared to the parent peptide, D-amino acid substitution could induce changes in this profile independent of its possible effect on conformation.

5. *Arginine vasopressin (AVP)*

The neurohypophyseal hormone AVP has the following primary structure

$$\text{H}-\overset{1}{\text{Cys}}-\overset{2}{\text{Tyr}}-\overset{3}{\text{Phe}}-\overset{4}{\text{Gln}}-\overset{5}{\text{Asn}}-\overset{6}{\text{Cys}}-\overset{7}{\text{Pro}}-\overset{8}{\text{Arg}}-\overset{9}{\text{Gly}}-\text{NH}_2$$

Central diabetes insipidus is a form of diabetes that is of central origin and vasopressin-sensitive. It is thought to be caused by a failure to maximally release endogenous AVP. Since AVP possesses both antidiuretic and vasoconstrictor properties, treatment of central diabetes insipidus with exogenous AVP involves an increased risk of hypertension. An analog of AVP has been developped that is virtually devoid of vascular properties, whereas its antidiuretic activity is several times higher as compared to that of the natural hormone. Moreover, its duration of action is considerably increased and the analog can be administered by intranasally. The analog is des-1-amino [D-Arg^8] vasopressin and this compound is widely and effectively used in clinical practice for the treatment of diabetes insipidus. The des-1-amino-vasopressin has similar vasoconstrictor but about 4× increased antidiuretic activity as compared to AVP. [D-Arg^8] vasopressin has about 1% of vasoconstrictor and about 33% of antidiuretic activity as compared to AVP. It can thus be concluded that substitution of the arginine residue in position-8 by its D-isomer alone decreases preferentially vasopressor activity[13].

Conclusions

The examples illustrate that D-amino acid substitutions involve changes in conformation and metabolism, the latter concerning overall metabolic stability of the parent peptide as well as accumulation of preferred (stable) intermediate peptides. It depends on the peptide(s) involved which of these changes dominate. The changes principally affect biological activity profiles, potency and durations of action.

As structure-activity relationships in the peptide field are complex, a rational approach in synthesizing clinically applicable D-amino acid substituted peptides is still far off, but sophisticated trial and error approaches have shown valuable and the outlooks for the future appear promising.

References

1. M. Broccardo, V. Erspamer, G. Falconieri-Erspamer, G. Improta, G. Linari, P. Melchiorri and P.C. Montecucchi. Brit. J. Pharmacol. *73*, 625-631 (1981).
2. A.M. Bool, G.H. Gray II, M.E. Hadley, C.B. Heward, V.J. Hruby, T.K. Sawyer and Y.C.S. Yang. J. Endocr. *88*, 57-65 (1981).
3. O. Hechter, T. Kato, S.H. Nakagawa, F. Yang and G. Flouret. Proc. Nat. Acad. Sci. USA *72*, 563-566 (1975).
4. W. Vale, C. Rivier and M. Brown. Ann. Rev. Physiol. *39*, 477-482 (1977).
5. R.N. Clayton, J.P. Harwood and K.J. Catt. Nature *282*, 90-92 (1979).
6. J.J. Reeves, C. Séguin, F.A. Lefebre, P.A. Kelly and F. Labrie. Proc. Nat. Acad. Sci. USA *77*, 5567-5571 (1980).
7. R.F. Casper and S.S.C. Yen. Science *205*, 408-410 (1979).
8. A.J.W. Hsueh and N.C. Ling. Life Sci. *25*, 1223-1230 (1979).
9. J.S. Morley. Ann. Rev. Pharmacol. Toxicol. *20*, 81-110 (1980).
10. H.M. Greven and D. de Wied. In: Hormones and the Brain p. 115-126 (1980), D. de Wied and P.A. van Keep (eds.). Falcon House: MTP.
11. A. Witter, H.M. Greven and D. de Wied, J. Pharmacol. Exp. Ther. *193*, 853-860 (1975).
12. M. Fekete and D. de Wied. Eur. J. Pharmacol. (1982), in press.
13. W.H. Sawyer, M. Acosta, L. Baláspiri, J. Judd and M. Manning. Endocrinology *94*, 1106-1115 (1974).

Stereoselectivity in Various Drug Fields

P.B.M.W.M. Timmermans

Abstract

Examples are given illustrating the influence of stereochemistry on the pharmacological profiles of various drug molecules. The dependence of specific biological properties on the precise geometrical configuration of double bonds is well demonstrated by the insect pheromone, bombykol. The most active isomer out of the four possibilities outstripped the next most active one by a factor of 10^9. Alteration of the stereochemistry of the angiotensin-converting enzyme inhibitor, captopril, leads to substantial variations in activity. Structure-activity studies on chiral GABA agonists and GABA-uptake blockers of known configuration have provided information on the molecular aspects of GABA-receptor interactions. Different activities of the two enantiomers of the calcium entry blockers verapamil, D 600 and nifedipine and derivatives probably contribute to the overall effects of these therapeutically important drugs. The (+)-R-isomer of the anaesthetic agent etomidate shows all the pharmacological activity of the racemic mixture. Both enantiomers of indacrinone have uricosuric activity, but the diuretic activity resides predominantly in the (-)-enantiomer. The uricosuric properties of indacrinone can be enhanced by manipulation of the enantiomer ratio. The asymmetry of only two of the four chiral carbon atoms present in quinine are important for antimalarial activity. The erythro form, quinine and quinidine are active; the threo forms are inactive. Similarly, the antiamebic activity is limited to the levorotatory from of emetine only. Pronounced central hypotensive activity is found for the erythro mixture of R 28935, a benzodioxane-hydroxyethylpiperidine derivative. The threo form, R 29814, is much less potent. In contrast, R 28935 and R 29814 seem to label identical displaceable binding sites in the brain with comparable affinity. The antidepressant activity of mianserin probably resides in the (+)-S-enantiomer. However, this stereoselectivity is hardly reflected by the α-adrenoceptor and serotonin receptor blocking activities of the individual (+)-S- and (-)-R-isomers.

Introduction

Optical isomer pairs possess unique physiochemical properties which permit them to serve an important role in pharmacological investigations. Enantiomers exhibit identical physical properties and show differences only at the molecular level, where they interact with selective systems which can discriminate between their particular stereochemical orientations. Receptor stereoselectivity has become an operational criterion of receptor identification. The overall physiological effects produced by stereoisomeric forms of a given structure are frequently pronounced. It is, however, often the case that a precise explanation of the observed effects in molecular terms is impossible at present. In the following, the impact of stereochemistry on the selectivity and pharmacological profile of biological active molecules will be illustrated with examples out of various drug fields. The examples are chosen in such a way to cover a wide range of biological responses.

Bombykol

Some interest has been concentrated on olfactory effects in insects especially after the extensive studies on the isolation, structure and function of sex attractants[1-3]. The dependence of specific biological properties on the precise geometrical configuration of double bonds is well illustrated by the insect pheromone Bombykol, the attracting substance of the silkworm moth, *Bombyx Mori,* which was found to have the structure 10-*trans*-12-*cis*-hexadecadienol (**1**). The isomers of this compound have been synthesized[1]. The action on the receptors required a high specificity in structure. As shown in Table 1, the threshold values for the three stereoisomers of naturally occurring Bombykol were found to be 10^9-10^{13} times higher. The increased knowledge about sex attractants in insects has been of considerable practical importance in view of their application in insect control.

$$CH_3CH_2CH_2-\underset{12}{CH}=CH-\underset{11}{CH}=\underset{10}{CH}-(CH_2)_8CH_2OH$$

1

Table 1. *Biological Activities of Bombykol Stereoisomers*[a]

	Activity Units (μg/ml)
10-*Trans*-12-*cis* (**1**)	10^{-12}
10-*Cis*-12-*trans*	10^{-3}
10-*Cis*-12-*cis*	1
10-*Trans*-12-*trans*	10

[a] From ref. 1.

Converting enzyme inhibitors

Encouraged by the apparent specificity of succinyl-L-proline (**2**) as an inhibitor of angiotensin converting enzyme (ACE), scientists of the Squibb Institute embarked upon structure-activity studies that eventually led to the synthesis of hundreds of analogues designed to test their hypothesis and to improve the modest activity observed with this prototype[4-9]. Five postulated enzyme binding interactions of succinyl-L-proline (**2**) and its analogues formed the basis for these highly directed structure-activity studies (Table 2): side chain interactions with subsites S_1' and S_2', ionic interaction of the amino acid carboxylate with a positively charged residue of the enzyme, hydrogen bonding of the amide carbonyl with the putative donor X-H and interaction of the succinyl carboxylate or alternate structure with the zinc ion of the enzyme.

In agreement with the postulated mode of binding of succinyl-L-proline (**2**) to the active site of the converting enzym, introduction of a substituent in the 2-position of the succinyl group enhanced inhibitory activity, but only if its steric configuration

Table 2. *Structures and Activities of Important Intermediates* (**2, 3** *and* **4**) *in the Design of Captopril* (**5**).
The diagrammic model of the active site of angiotensin converting enzyme is explained in the text. As demonstrated by the structures (**6**), (**7**) *and* (**8**), *alteration of the stereochemistry of the side-chain leads to substantial variation in activity.* I_{50}*: concentration producing 50% in vitro inhibition of rabbit lung ACE*[6,7,10-13].

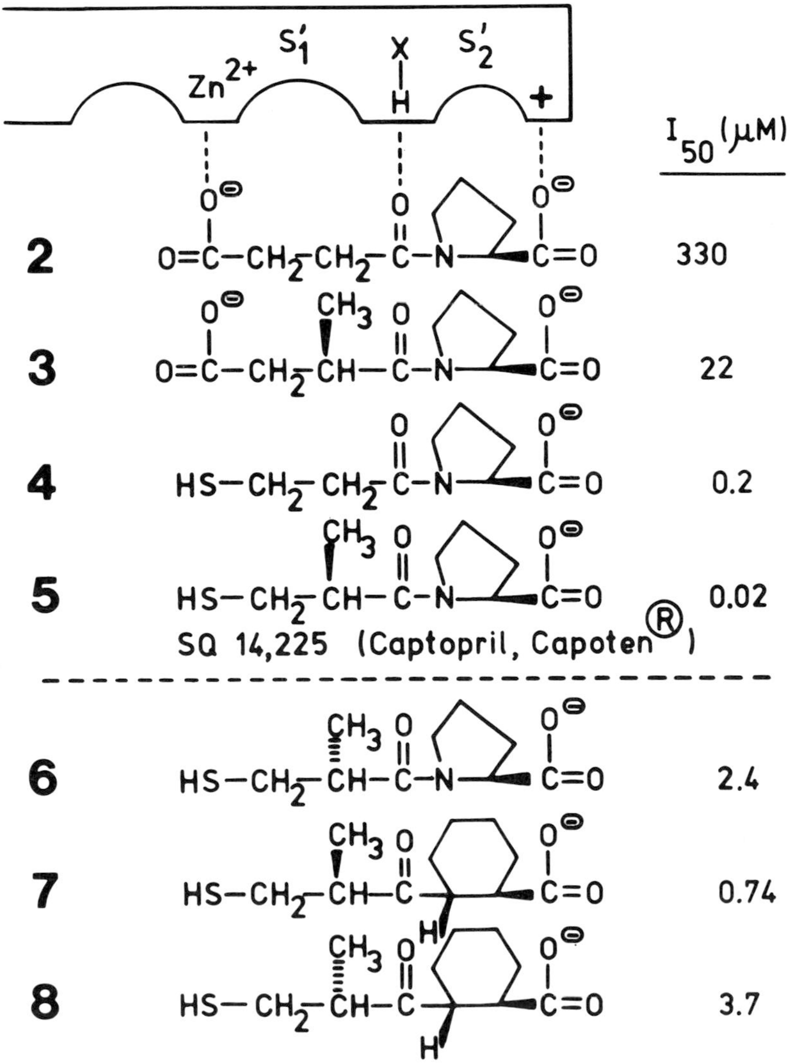

S_1' X S_2'
Zn^{2+} H +

	Structure	I_{50} (μM)
2	$O^{\ominus}$–O=C–CH_2–CH_2–C(=O)–N(proline)–C(=O)$O^{\ominus}$	330
3	$O^{\ominus}$–O=C–CH_2–CH(CH_3)–C(=O)–N(proline)–C(=O)$O^{\ominus}$	22
4	HS–CH_2–CH_2–C(=O)–N(proline)–C(=O)$O^{\ominus}$	0.2
5	HS–CH_2–CH(CH_3)–C(=O)–N(proline)–C(=O)$O^{\ominus}$ SQ 14,225 (Captopril, Capoten®)	0.02
6	HS–CH_2–CH(CH_3)–C(=O)–N(proline)–C(=O)$O^{\ominus}$	2.4
7	HS–CH_2–CH(CH_3)–C(=O)–(cyclohexane, H)–C(=O)$O^{\ominus}$	0.74
8	HS–CH_2–CH(CH_3)–C(=O)–(cyclohexane, H)–C(=O)$O^{\ominus}$	3.7

mimicked that of the penultimate L-amino acid residue of a substrate that normally interacts with subside S_1' of the enzyme. One of the most effective of such substituents has been the methyl group; incorporation of this side chain yields D-2-methylsuccinyl-L-proline (**3**, Table 2) which was 15 to 20 times more inhibitory than succinyl-L-proline (**2**). Compound (**3**) was the first of this series to be shown to be active *in vivo* after oral administration[4]. With regard to enhancing inhibitory activity, the most important of the five postulated interactions between compounds such as (**3**) and the active site of angiotensin converting enzyme turned out to be the binding of the succinyl carboxylate to the zinc ion of the enzyme. When this group was replaced by sulfhydryl which appeared from existing data on zinc complexes to be a better ligand for zinc than carboxylate, the resulting compounds (**4** and **5**) were 1000 times more inhibitory than their carboxylic acid precursors (**2** and **3**, Table 2). Captopril (**5**) is a very potent competitive inhibitor of ACE. Its specific binding to the active site of this enzyme has not been substantially bettered in testing hundreds of carefully selected analogues. As visualized in Table 2, the inhibitory activity of captopril (**5**) and analogues heavily depends on the configuration of the side-chain. Alteration of the stereochemistry of the α-methyl group profoundly affects potency.

Gabaergic drugs

The role of γ-aminobutyric acid (GABA) as an inhibitory neurotransmitter in the mammalian central nervous system is well established and has been the subject of numerous comprehensive reviews[14-19]. The present knowledge of the location and function of GABA receptors in GABAergic synapses are illustrated in Fig. 1. The presynaptic autoreceptor (*A*) is assumed to regulate GABA release. GABA-T catalyzes the conversion of GABA into succinic semialdehyde (SSA). GABA uptake processes have been identified in glia cells (*B*) and in nerve terminals (*C*). Postsynaptic GABA receptors are coupled to Cl^- ionophores. Stimulation results into an increased membrane permeability to Cl^-. High (*D*)- and low (*E*)-affinity binding sites for GABA on the postsynaptic receptor appear to exist.

The processes and functions illustrated in Fig. 1 are particularly stereoselective and susceptible to pharmacological manipulation as evident from the following examples. Compounds with potent and specific actions on different functions of

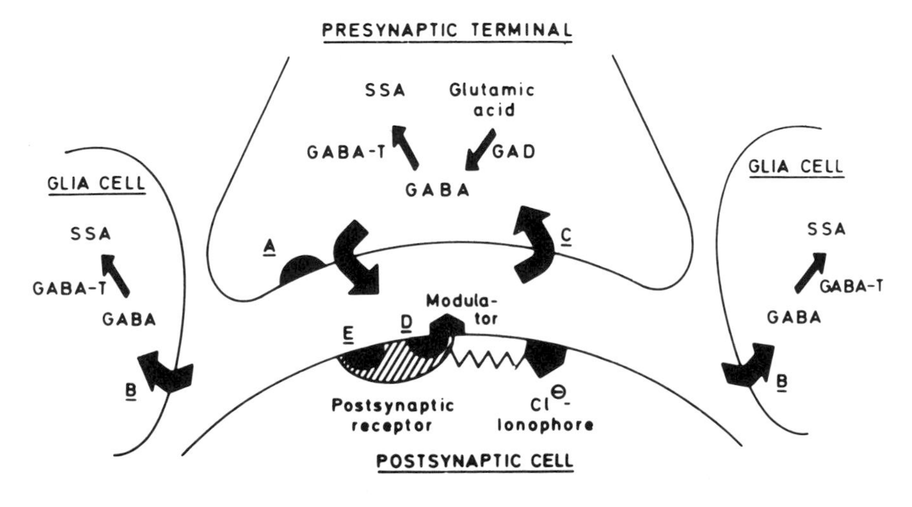

Figure 1. *GABAergic synapse. For explanation see text. From ref. 31.*

GABAergic neurones are vital tools for studies of the GABA system at the cellular and molecular level. Furthermore, these model compounds represent important steps in the attempts to develop therapeutically useful GABAergic drugs. The GABA receptors exhibit stereoselectivity with respect to antagonists, like bicuculline (**9**). Thus (–)-(1R,9S)-bicuculline is much weaker than (+)-(1S,9R)-bicuculline as an inhibitor of ^{3}H-bicuculline methiodide binding[20,21] and (+)-(1S,9R)-narcotine (**10**) is much more potent than its optical isomers[22,23].

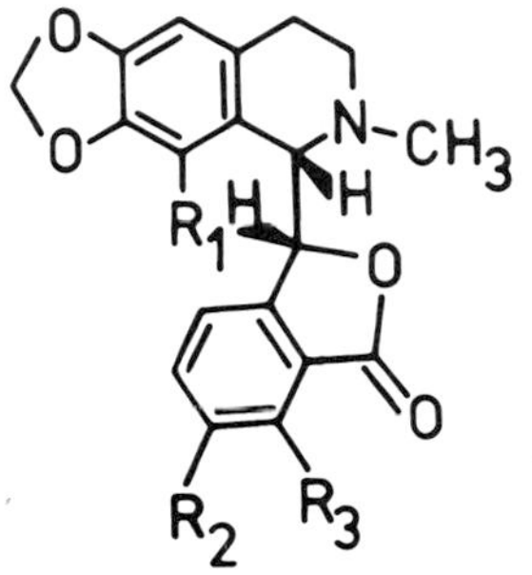

9 $R_1 = H$; $R_2,R_3 = OCH_2O$

10 $R_1 = R_2 = R_3 = OCH_3$

Whereas baclofen (**11**) is a very weak and nonstereoselective inhibitor of the binding of ^{3}H-GABA to rat brain membranes[24], its (–)-isomer is more than 100 times more effective than the (+)-isomer to behave as an agonist of so-called prejunctional (bicuculline-insensitive) GABA receptors on sympathetic nerve terminals inhibiting transmitter release[25].

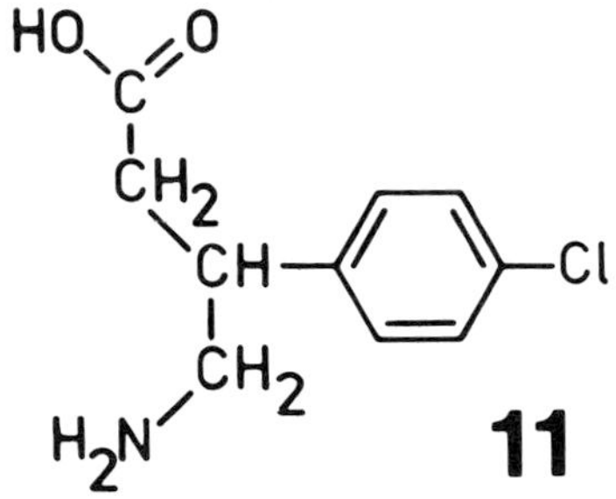

11

Structure-activity studies on chiral GABA agonists of known configuration have provided information about the molecular aspects of GABA receptor interactions. (S)- and (R)-4-Methyl-GABA (**12**) are equipotent as inhibitors of GABA binding, whereas the (S)-forms of the more rigid GABA analogues (**13**) and (**14**) are 20-30 times more effective than the (R)-isomers[26,27]. Similarly, (1S,3S)-(**15**) is a more powerful GABA agonist than its optical antipode[28]. Consequently, the mode of interaction of the flexible GABA analogue 4-methyl-GABA (**12**) with the receptor must be different from that of the conformationally restricted congeners concerned. These differences may be attributed to different degrees of intimacy of binding of the respective compounds to the receptor site. An alternative interpretation of these stereostructure-activity studies is that 4-methyl-GABA (**12**) interacts with a binding site of the GABA receptor different from that activated by the methylated derivatives of muscimol (**14**) and trans-(**13**). In fact, two binding sites for GABA showing different binding characteristics have been detected in synaptic membranes treated with detergent[27,29,30]. Thus the possibility may be considered that different confor-

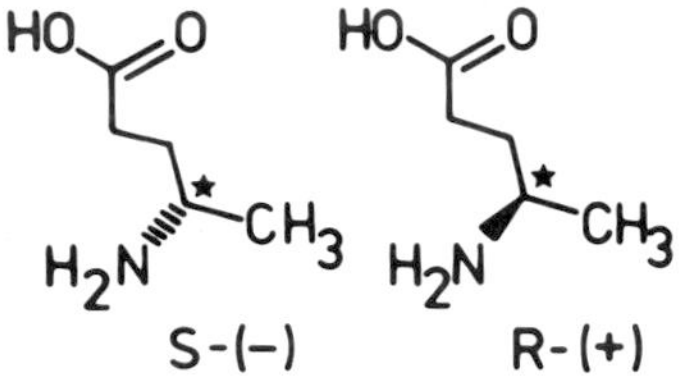

12: S-(−) 4.7; R-(+) 5.0

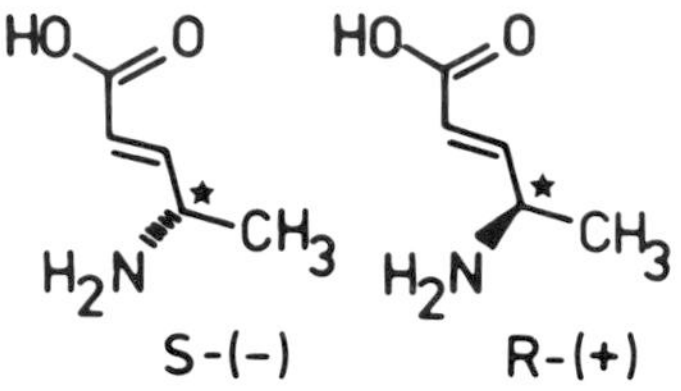

Inhibition of GABA-binding (μM)

13: S-(−) 4.1; R-(+) 148

14: S-(−) 0.6; R-(+) 19

15: 1S,3S; 1R,3R

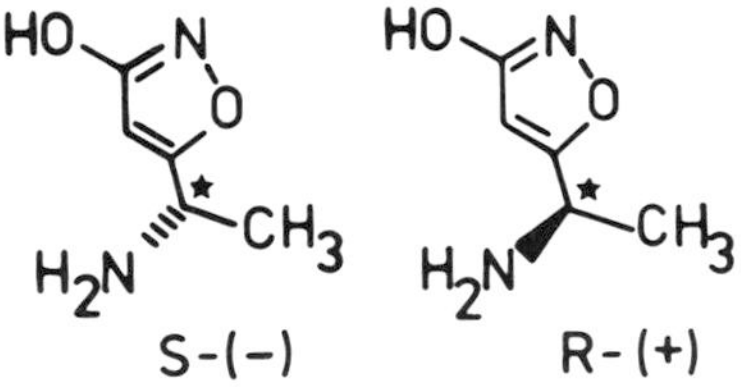

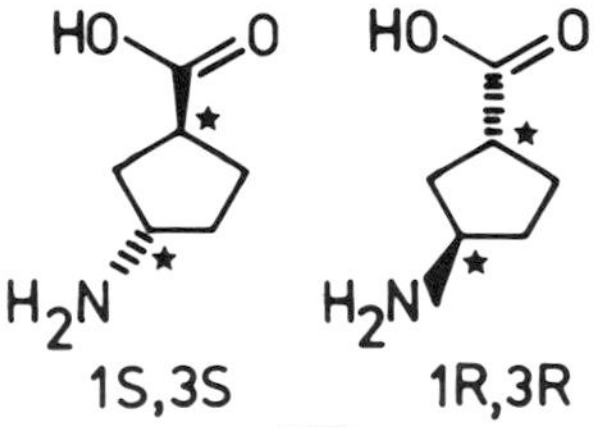

mational modes of GABA interact with different GABA receptor sites[31]. With respect to inhibition of neuronal and glial GABA uptake, the (R)-isomers of **(12)** and **(13)** are, however, more potent than the corresponding (S)-forms[27], emphasizing the different 'active conformation' of GABA on the receptor and transport carrier.

Differences in terms of structural requirements have also been detected for the GABAergic molecules transported via the neuronal and glial transport systems[32]. As listed in Table 3, for 4-methyl-GABA **(12)** and *trans*-4-amino-crotonic acid **(13)**, the congeners in the R-configuration are more potent in inhibiting GABA-uptake into cultered astrocytes (glial transport) and into mini-slices (neuronal transport). It should, however, be noted that R-*trans*-4-methyl-4-amino-crotonic acid **(13A)** seems to be a more potent inhibitor of the neuronal GABA transport system than of the glial system. The stereoselectivity exhibited by the difference in inhibitory potency of the uptake systems between R and S-forms of the *trans*-4-methyl-4-amino-crotonic acid **(13)** is exactly opposite to the selectivity reported for the GABA receptor (see above). In addition, R-(−)-Nipecotic acid **(16B)** is much more potent than the S-(+)-isomer. The IC_{50} for (+)-nipecotic acid **(16A)** is, however, much higher for the glial transport than for the neuronal transport. This could mean that the former system is more stereoselective than the latter. Thus the (+)-isomer **(16A)** can be used as a specific blocker of neuronal GABA transport. Its ethyl ester (a prodrug) is of potential usefulness in the treatment of diseases, like Huntington's chorea.

Calcium entry blockers

Calcium entry blockers are a rather heterogenous class of drugs recognized as useful antiarrhythmic agents for about a decade. More recently, their vascular effects have been shown to be of potential therapeutic and pharmacological interest, especially in the treatment of angina pectoris involving coronary arterial spasm and possibly also of hypertension[33–36]. The structures belonging to the group of calcium

Table 3. *Stereoselectivity in the inhibition of GABA uptake in astrocytes (glial transport) and cortical mini-slices (neuronal uptake)*[a]

	IC_{50} (μM)	
	Astrocytes	Mini-slices
12 A	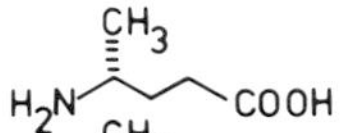 120	200
12 B	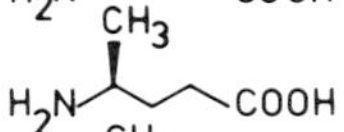 1000	750
13 A	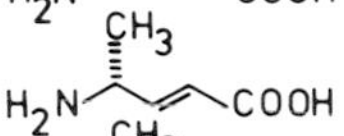 500	160
13 B	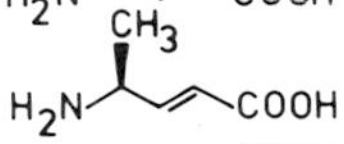 >5000	>5000
16 A	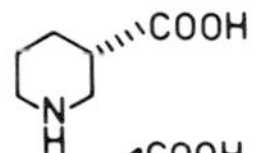 1000	30
16 B	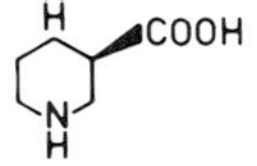 30	5

[a] From ref. 32.

entry blockers share one common property, that is the inhibition of the so-called slow inward current carried by calcium ions through voltage- or receptor-operated membrane channels in various cardiac and smooth muscle tissues. Some therapeutically applied and experimental calcium slow channel blockers, like D 600 (**17**), verapamil (**18**), nifedipine (**19**), nimodipine (**20**) and nitrendipine (**21**) are depicted below.

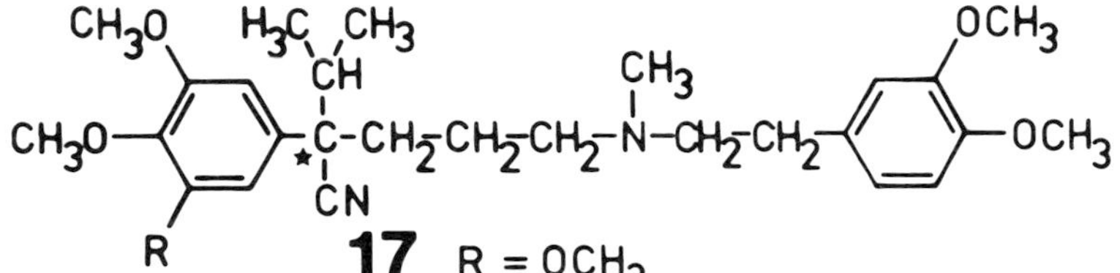

17 R = OCH_3
18 R = H

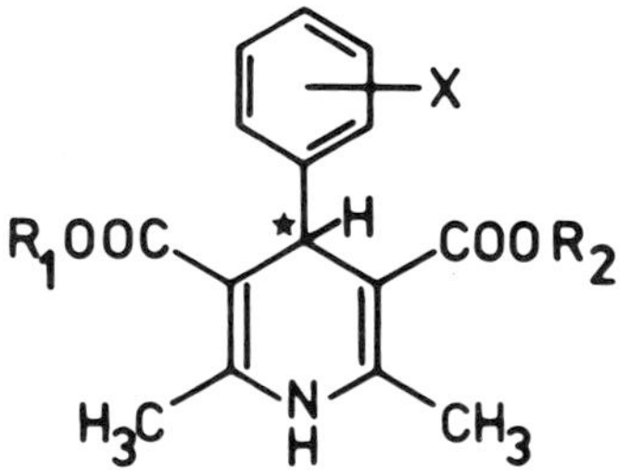

	R_1	R_2	X
19	CH_3	CH_3	2-NO_2
20	i-C_3H_7	CH_2-CH_2-OCH_3	3-NO_2
21	CH_3	C_2H_5	3-NO_2

Derivatives of verapamil (**18**) and nifedipine (**19**) possess one asymmetric carbon atom and they therefore represent a racemic mixture of (+)- and (–)-enantiomers. It has been suggested that different activities of the two enantiomers may contribute to the several pharmacological activities of these drugs. Accordingly, studies have been undertaken to compare potencies of the individual optical isomers.

Stereoselectivity of antagonism by D 600 (**17**) of smooth muscle responses induced by agonists (noradrenaline or *cis*-2-methyl-4-dimethylaminomethyl-1,3-dioxolane methiodide) or by K^+-depolarization has been observed[37]. Some data have been listed in Table 4. (–)-D 600 was the more potent stereoisomer, although the stereoselectivity was not constant and more pronounced with the tonic component of contraction. A similar stereoselectivity was seen in the inhibition of the Ca^{2+}/K^+ response (Table 4). An absence of stereoselectivity by the enantiomers of D 600 was observed against radioligand binding to muscarinic (^{3}H-QNB) and α_1-adrenergic (^{3}H-WB-4101) receptors (Table 4).

Table 4. *Activity ratios of (–)- and (+)-D 600* (**17**) *with respect to inhibition of smooth muscle responses and binding of radioligands*[a]

		Ratio (–) : (+)
Guinea-pig ileum		
CD[b]-induced	Phasic	2.5
	Tonic	180
Ca^{2+}/K^+-induced		10
Rat vas deferens		
NA[c]-induced	Phasic	5
	Tonic	15
Ca^{2+}/K^+-induced		40
Rat portal vein		
K^+-induced	Phasic	8.4
	Tonic	16.7
Radioligand binding		
^{3}H-QNB[d] to guinea-pig ileum		0.8
^{3}H-WB-4101[e] to rat vas deferens		1.1

[a] From ref. 37.
[b] CD = *cis*-2-methyl-4-dimethylaminomethyl-1,3-dioxolane methiodide
[c] NA = Noradrenaline
[d] QNB = Quinuclidine benzylate
[e] WB-4101 = 2-[(2′,6′-dimethoxyphenoxyethyl)-aminomethyl]-1,4-benzodioxane

As shown in Table 5, (–)-verapamil also showed enhanced activity over (+)-verapamil to act as a calcium entry blocker with respect to α_2-adrenoceptor-mediated vasoconstriction[38], a process which is carried by the influx of calcium ions[39–42]. In accordance to the α_1-adrenoceptor binding affinity of (+)- and (–)-D 600, (+)- and (–)-verapamil also showed similar affinities to this binding site when identified with ^{3}H-prazosin. On the other hand, some stereoselectivity of (+)- and (–)-verapamil with respect to ^{3}H-clonidine displaceable binding sites (α_2-adrenoceptors) was found[38].

Table 5. *Activity ratios of (-)- and (+)-verapamil (**18**) with respect to inhibition of smooth muscle responses and binding of radioligands*[a]

	Ratio (-) : (+)
Pithed rat; α_2-adrenoceptor-induced vasoconstriction	7
^{3}H-Prazosin to rat brain	0.6
^{3}H-Clonidine to rat brain	7

[a] From ref. 38.

The optical isomers of the 1,4-dihydropyridine derivative, nimodipine (**20**), a potent blocker of calcium movements in vascular smooth muscle, also showed the stereoselectivity of the interaction of this kind of drugs in or near the calcium channel[43] (see Table 6).

Table 6. *Activity ratios of (-)- and (+)-nimodipine* (**20**) *with respect to inhibition of smooth muscle responses induced by* K^+[a]

	Ratio (-) : (+)
Rabbit aorta	5.3
Rabbit basilar artery	7.5
Rabbit saphenous artery	4.3

[a] From ref. 43.

The enhanced activity of the (-)-enantiomers over the (+)-isomers of D 600 (**17**), verapamil (**18**) and nimodipine (**20**), to act as slow calcium channel blockers is expressed by the observation that the (-)-stereoisomers are more potent negative inotropic agents and hypotensive drugs. It is interesting to add that in equieffective doses in producing coronary vasodilatation in dogs, (+)-verapamil has been reported to be far less cardiodepressive than the (-)-isomer, so that the former would be a more desirable antianginal drug for reasons of safety[44].

The tritiated calcium entry blockers nimodipine (**20**) and nitrendipine (**21**), potent analogues of nifedipine (**19**), bind in a reversible and saturable manner to partially purified guinea-pig or rat heart, kidney, lung and brain membranes in the nanomolar range[45-47]. The displaceable binding sites are stereoselective and discriminate between (+)- and (-)-nitrendipine, the (-)-stereoisomer being much more potent than the (+)-isomer[45]. It is intriguing that verapamil and D 600 were found to be at least several hundred times weaker than nifedipine in competing for ^{3}H-nitrendipine and ^{3}H-nimodipine binding sites[45,46]. This finding strongly suggests that ^{3}H-nimodipine and ^{3}H-nitrendipine label sites which differ from those identified by verapamil and its methoxy derivative D 600. However, it is also possible that this binding site is only accessible for the dihydropyridine structure, so that it is rather a structure-recognition site not necessarily connected to pharmacologically relevant binding sites at calcium channels.

Etomidate

Etomidate (**22**) is a potent, short-acting, non-barbiturate intravenous hypnotic in animals[48,49] and man[50,51]. The drug is a pure hypnotic and has no analgesic activity. It has a rapid onset of action and induces hypnosis within one minute. R-(+)-Etomidate possesses the hypnotic activity, the S-(–)-isomer being devoid of hypnotic potency[52]. Although the mechanism of etomidate's hypnotic action is not known in detail, the stereoselectivity observed indicates specificity of the brain 'receptors' responsible for this pharmacological effect. Stereoselectivity has also been described for the hypnotic action of the enantiomers of hexobarbital[53].

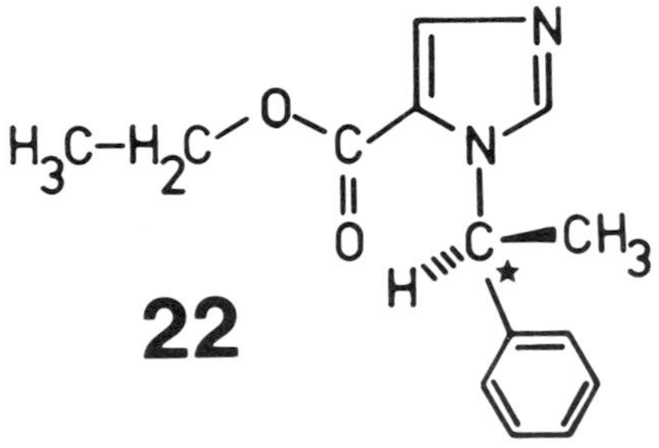

$H_3C-H_2C-O-C(=O)-$ … **22**

Indacrinone

Racemic indacrinone (**23**) is a potent, high-ceiling, relatively long acting diuretic[54]. Both enantiomers have uricosuric activity, but the diuretic activity resides predominantly in the (–)-enantiomer[55]. However, usual therapeutic doses of racemic indacrinone (**23**) have only transient (4-6 hr) uricosuric activity[54,56], so that hyperuricemia occurs as with most diuretic agents[57]. The relatively high uricosuric/natriuretic potency of the (+)-enantiomer has been used to improve the pharmacological profile of the drug[58]. Increasing the ratio of (+)-indacrinone to the (–)-isomer

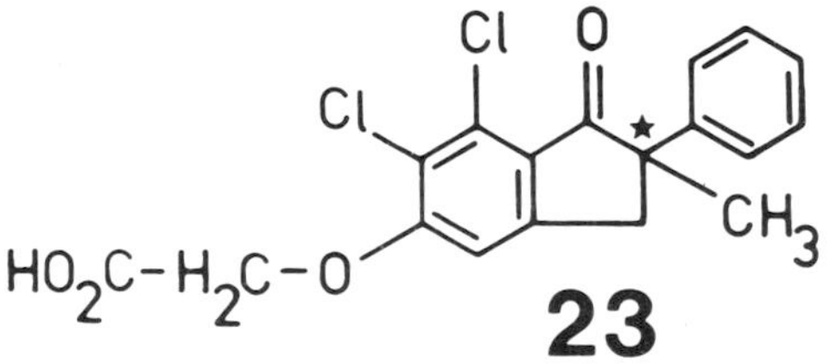

HO_2C-H_2C-O- … **23**

would be expected to prevent or reverse the hyperuricemic effect of the racemate without inducing excessive natriuresis. For this purpose healthy men with normal renal, hepatic and cardiac functions were allocated to indacrinone (–)/(+) ratios (mg/mg) of 10/0, 10/10, 10/20, 10/40 and 10/80, given once daily at 0900 hr for 7 consecutive days. There was a direct correlation between increasing dose of (+)-indacrinone and decreasing plasma uric acid ($p < 0.01$) for all measurements from day 1 (evening) through day 7 (evening, see Fig. 2). The falls in plasma uric acid were particularly evident with the 40- and 80-mg (+)-indacrinone doses. (–)/(+)-Indacrinone 10/40 was approximately isourecemic and 10/80 produced a moderate lowering of plasma uric acid. This study represents the first attempt to improve the therapeutic effects of a drug by manipulation of the enantiomer ratio.

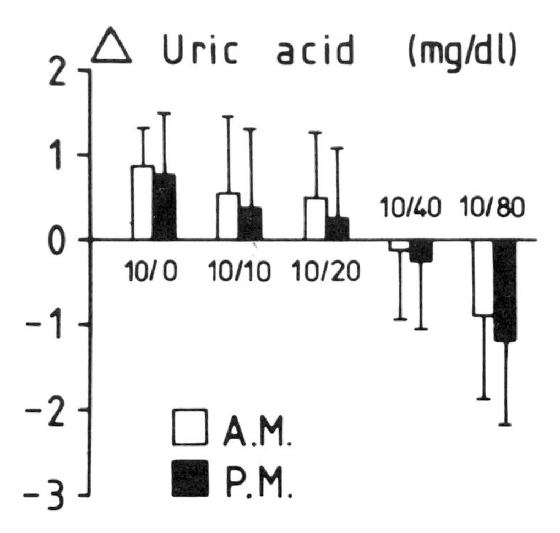

Figure 2. *Change in plasma uric acid ($\bar{x} \pm SD$) of healthy men at day 7 of a consecutive treatment with oral indacrinone* (**23**) *in various (+)/(-) ratios (10/0 to 10/80, mg/mg) given once daily (n = 18, morning; n = 17, evening). Modified after ref. 58.*

Antimalarial agents

Quinine (**24**) is one of the best known amino alcohols belonging to the class of antimalarial agents. The structure of quinine with its four asymmetric centers at C_3, C_4, C_8 and C_9, has been examined thoroughly to establish the parts of the molecule responsible for the antimalarial activity[59,60]. It has been concluded that the asymmetry at positions 3 and 4 is not essential for antimalarial activity, since activity is still found for molecules lacking these two asymmetric centers[59,61]. However, the configurations at positions 8 and 9 affect the juxtaposition of the hydroxy group and the nonaromatic nitrogen atom, a relationship that seems to be associated with antimalarial activity. Because of the presence of these two asymmetric centers, two erythro forms, quinine (**24**) and quinidine (**25**) and two threo forms, 9-epiquinine (**26**) and 9-epiquinidine (**27**) exist. It is likely that the bulkiness of the quinuclidine

(8S,9R) **24**

(8R,9S) **25**

(8S,9S) **26**

(8R,9R) **27**

moiety in the threo forms prevent the juxtaposition of the hydroxy group and the nitrogen atom of the quinuclidine nucleus that is necessary for activity. Support for this view is obtained when the rigid quinuclidine nucleus is replaced with the more flexible piperidyl group, since both the erythro and threo forms were found to be slightly active[59]. By using Dreiding models the respective orientations of the molecules demonstrated that for antimalarial activity, the distance between the oxygen and the nonaromatic nitrogen should be about 3 Å[62]. It has been postulated that a compound will be inactive when this distance is greater in the compound's preferred conformation[63]. Apparently, in all active amino alcohol-type antimalarials the orientation of the hydroxy hydrogen and the amino nitrogen should be such that hydrogen bonding is possible.

Emetine, an antiamebic agent

The gross structure of emetine was finally elucidated in 1949[64,65] and the structure (**28**) is now accepted as the configuration of the natural alkaloid (–)-emetine[66–69]. A stereochemically favorable synthesis of (–)-emetine has been achieved[70] and is now used on a commercial scale. Since emetine has four asymmetric centers, a number of stereoisomers are possible. However, the antiamebic activity of (–)-emetine is highly stereospecific, since all the stereoisomers of emetine tested for amebicidal activity *in vitro* or *in vivo* are less active than the natural alkaloid[71–74]. The modes of action of (–)-emetine and related compounds have been studied in some detail and it appears that all the active compounds inhibit protein synthesis in certain mammalian and other cells[75–77]. The observation has been made that (–)-emetine and certain other alkaloids show configurational similarities to (–)-cycloheximide (**29**) and the glutarimide antibiotics and display similar effects on protein synthesis[78]. This led to the proposal that structure (**30**) contains the topochemical requirements for amebicidal activity and for the inhibition of protein synthesis[78].

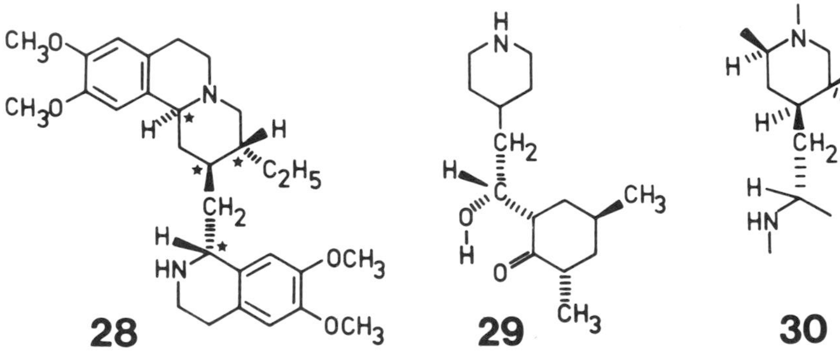

CH_3O CH_3O N H H C_2H_5 CH_2 H HN OCH_3 OCH_3

28

H N CH_2 H C O H CH_3 O CH_3

29

N H H R H CH_2 H HN

30

Benzodioxane-hydroxyethylpiperidine derivatives possessing central hypotensive activity

The erythro mixture R 28935 (**31**), a benzodioxane derivative, developed from the neuroleptic agent pimozide, is a potent hypotensive drug in various animal species[79–85]. The hypotensive effect of R 28935 is of central nervous origin and stereospecific, since the threo form R 29814 (**32**) is much less potent[80,82,83,85,86]. Following intravenous administration to anaesthetized rats (Fig. 3) as well as vertebral arterial infusion into anaesthetized cats (Fig. 4), the erythro mixture R 28935 (**31**) was found about 10-30 times more potent in lowering mean arterial pressure than threo R 29814 (**32**).

31 Erythro : R 28935
32 Threo : R 29814

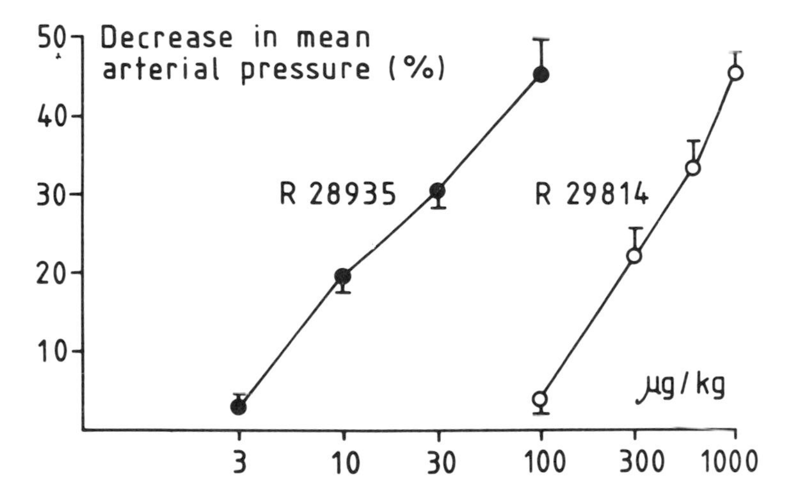

Figure 3. *Log dose-response relationships of R 28935* (**31**) *and R 29814* (**32**) *with respect to the maximal decrease in mean arterial pressure of anaesthetized rats following intravenous injections. Modified after ref. 86.*

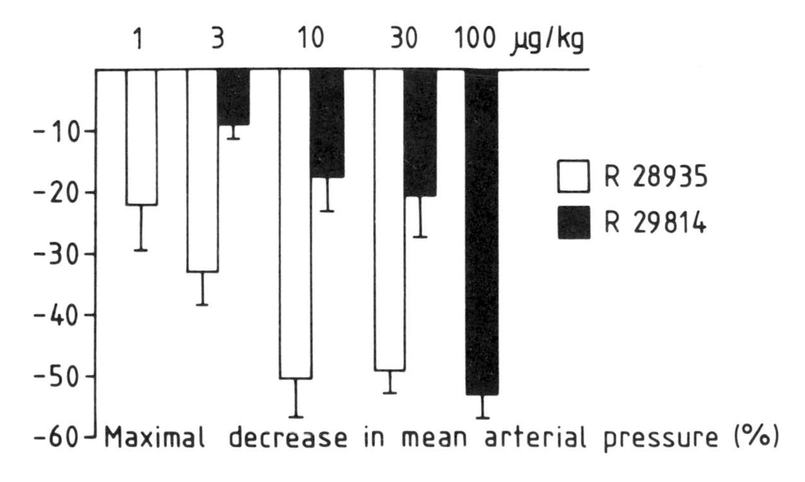

Figure 4. *Hypotensive effects of R 28935* (**31**) *and R 29814* (**32**) *after infusion into the vertebral artery of anaesthetized cats. Modified after ref. 80.*

The details of the central mode of action via which R 28935 (R 29814) induces hypotension is still unknown. At the time of the discovery of the pronounced hypotensive properties of R 28935, the involvement of central α-adrenoceptors has been considered unlikely, since the commonly known α-adrenoceptor blocking drugs, like yohimbine and piperoxan[80–82,84] or phentolamine and tolazoline[79] were ineffective in antagonizing the hypotension of R 28935. A rather surprising observation was made subsequently that the selective blocker of α_1-adrenoceptors, prazosin, proved able to attenuate the central hypotensive action of R 28935 already in very low doses[85,87,88]. The logical conclusion of these experiments that central α_1-adrenoceptors are the mutual sites of interaction is, however, not supported by the lack of α_1-adrenoceptor agonistic activity of R 28935[80–83,85]. In fact, the affinity of R 28935 for α_1-adrenoceptors is moderate and bears no relationship to hypotensive activity[85,86,89].

In the attempts to learn more details of the mechanism of action of this kind of drugs, it was studied whether specific binding sites for ^{3}H-R 28935 as well as ^{3}H-R 29814 could be identified in brain tissue and whether these sites were possible associated with the stereospecific receptor sites responsible for the initiation of the hypotensive effect.

Displaceable binding sites were identified for ^{3}H-R 28935 (9 Ci/mmol) in membranes from rat brain[90]. This binding was of high affinity, rapid and reversible with at least two components ($K_D = 0.25$ an 3 nM). The maximal number of specific binding sites amounted to 30 and 230 fmol/mg of protein, respectively. The displaceable binding of ^{3}H-R 28935 was effectively inhibited by its structurally related derivatives, but was hardly stereoselective, since R 29814 was only 3.5 times less active than R 28935 itself. Part of the binding only occurred to α_1-adrenoceptor-like binding sites. Furthermore, affinity for the sites identified by ^{3}H-R 28935 was weakly correlated with hypotensive activity. No other class of drugs investigated showed high affinity inhibition.

^{3}H-R 29814 (9 Ci/mmol) bound to homogenates of rat brain with one sarurable component with high affinity ($K_D = 1.1$ nM). The displaceable binding was rapid and reversible to a finite number of binding sites[91]. Similarly to ^{3}H-R 28935, the drug specificity of the ^{3}H-R 29814 binding sites was not associated with that of any known drug-recognition site. It also showed no stereoselectivity, the affinity of R 28935 and R 29814 being comparable. The specificity of the displaceable binding sites of ^{3}H-R 29814 corresponded to that of ^{3}H-R 28935, since a linear relationship could be derived between the affinities of various drugs for inhibiting ^{3}H-R 29814 and ^{3}H-R 28935 binding, as shown in Fig. 5.

The results of the binding studies suggest that ^{3}H-R 29814 labels identical binding sites as does ^{3}H-R 28935 with comparable affinity. However, it is then most unlikely that these sites are related to those responsible for initiating the hypotensive effect of this type of drugs in view of the great difference in pharmacological activity between both agents. Consequently, it may be concluded that both radioligands identify pharmacologically irrelevant structure-recognition sites only accessible to structurally closely related congeners. The details of the mechanism of the hypotensive action remain therefore still unknown.

Mianserin

Mianserin (**33**) is a clinically effective antidepressant (for review see ref 92). The drug has been described to possess presynaptic α-adrenoceptor blocking activity[93,94] in addition to noradrenaline (NA) uptake inhibiting properties[93,95,96]. However, blockade of NA-uptake has been suggested to represent the most significant peripheral effect of mianserin on noradrenergic transmission[97,98]. Both in the periphery

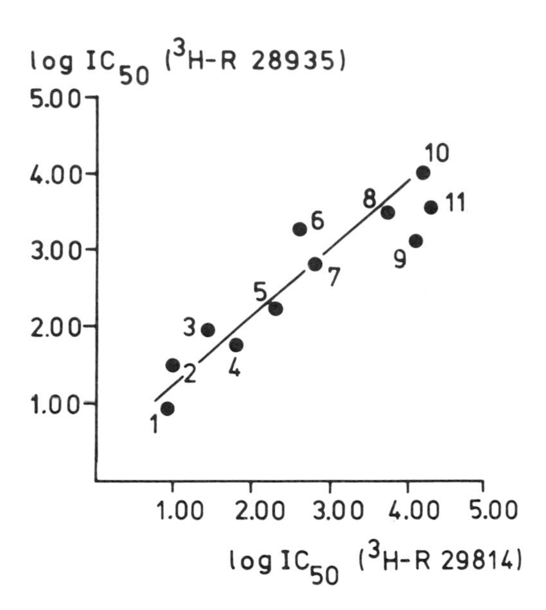

Figure 5. *Relationship between affinities of various drugs for the displaceable binding site of* ^{3}*H-R 28935 and* ^{3}*H-R 29814 in membranes from rat brain.* **Key:** *1 = R 28935; 2 = R 29814; 3 = R 36240; 4 = R 24315; 5 = R 31480; 6 = haloperidol; 7 = pimozide; 8 = yohimbine; 9 = methysergide; 10 = prazosin; 11 = (-)-propranolol. From ref. 91.*

and in the central nervous system mianserin displays potent antiserotonergic and antihistaminergic activities[99] (for review see ref. 100). On the other hand, mianserin lacks 5-hydroxytryptamine (5-HT) uptake blocking properties[95,96]. It is not well understood at present whether mianserin's antidepressant effect is related to its presynaptic α-blocking properties or is based upon the blockade of NA-uptake.

33

Mianserin (**33**) is the racemic mixture of the (+)-S- and the (-)-R-enantiomer. The (+)-isomer appears to be more active than the (-)-enantiomer in a number of tests considered to be predictive for antidepressant activity[101,102]. The antidepressant effect seems therefore stereospecific. A comparison between the various activities of the individual enantiomers may provide a more detailed insight into the features determining the antidepressant activity of this drug.

The various activities of (-)-R-, (+)-S- and racemic mianserin, involving effects on K^+-induced ^{3}H-NA release from rat cerebral cortex slices (= presynaptic α_2-adrenoceptor blocking activity *in vitro*), on NA- and 5-HT-uptake, on specific ^{3}H-dihydroergocryptine (DHE) (= affinity for α_1/α_2-adrenoceptors), ^{3}H-prazosin (α_1-adrenoceptors), ^{3}H-clonidine (α_2-adrenoceptors) and ^{3}H-($\pm$)-mianserin (5-HT_2-receptors) binding to rat brain membranes have been listed in Table 7.

Table 7. *Various pharmacological activities of racemic mianserin* (**33**) *and the individual (-)-R- and (+)-S-isomers*

	^{3}H-NA Release[a]	IC_{50} (M) Reuptake[b]		IC_{50} (M) Inhibition of Binding[c]			
		NA	5-HT	^{3}H-DHE	^{3}H-Prazosin	^{3}H-Clonidine	^{3}H-Mianserin
Racemate	+ + +[d]	3×10^{-8}	$> 10^{-5}$	–	3.8×10^{-7}	3.2×10^{-7}	6.7×10^{-9}
(-)-R	0	5×10^{-6}	$> 10^{-5}$	6×10^{-7}	2.2×10^{-6}	9.4×10^{-7}	3.5×10^{-8}
(+)-S	+ + +[d]	2×10^{-8}	$> 10^{-5}$	1×10^{-7}	2.7×10^{-7}	3.8×10^{-7}	3.7×10^{-9}

[a] From ref. 103.
[b] From ref. 102.
[c] From refs. 103-106.
[d] > 40% increase.

Fig. 6 shows the effect of mianserin and its enantiomers on the K^+-evoked release of ^{3}H-NA from rat cerebral cortex slices[103], which is obviously stereospecific; the activity resided in the (+)-isomer (also see Table 7). It should be remarked, however, that the antagonistic potency at presynaptic α_2-adrenoceptors which is to be deduced from these results, will be overestimated, since no correction has been made for inhibition of reuptake. In fact, (+)-S-mianserin is much more potent than (-)-R-mianserin to block NA-uptake. On the other hand, the potency of both enantiomers to interfere with 5-HT-uptake is very moderate. The difference in activity between (+)- and (-)-mianserin with respect to *in vitro* affinity for some receptor types identified by radioligands is greatest for 5-HT_2- and α_1-receptors labeled by ^{3}H-(±)-mianserin and ^{3}H-prazosin, respectively (Table 7), although this difference

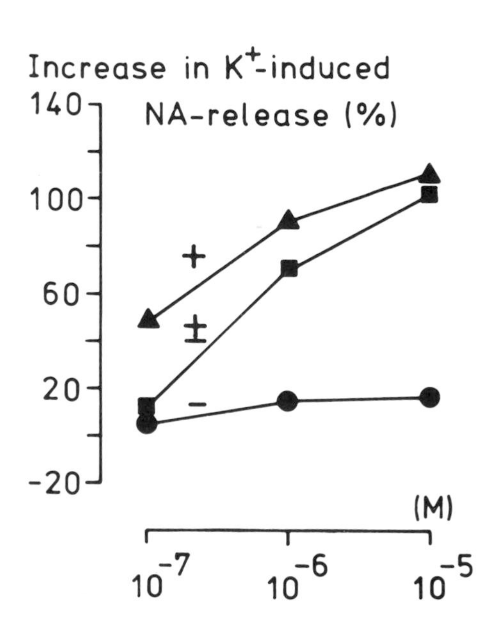

Figure 6. *Effects of (+)-, (-)- and (±)-mianserin* (**33**) *on the release of ^{3}H-noradrenaline (NA) from rat cerebral cortex slices induced by 26 mM K^+. Modified after ref. 103.*

is not particularly pronounced (about 10-fold). The stereoselectivity is even less regarding affinity for α_2-adrenoceptors. This contrasts with the differential activity of both enantiomers at presynaptic α_2-adrenoceptors (see above). However, functional antagonism *in vivo* of 5-HT_2-, α_1- and α_2-receptor-induced responses corresponds with the binding affinities of the individual enantiomers[104-106].

The data show that in all tests the (+)-S-enantiomer of mianserin, which presumably bears the antidepressant activity, shows highest potency. Consequently, it cannot be concluded from this comparison which property underlies the antidepressant activity. It is interesting to note that inhibition of NA-uptake seems more discriminative between (+)- and (–)-mianserin than blockade of α_2-adrenoceptors.

References

1. A. Butenandt and E. Hecker, *Angew. Chem. 73,* 350 (1961).
2. F. Crescitelli and T.A. Geismann, *Ann. Rev. Pharmacol. 2,* 143 (1962).
3. E. Hecker, *Umschau, 465,* 499 (1959).
4. M.A. Ondetti, B. Rubin and D.W. Cushman, *Science 196, 441 (1977).*
5. M.A. Ondetti, E.F. Sabo, K.A. Losee, H.S. Cheung, D.W. Cushman and B. Rubin, *In* 'Peptides-Proceedings of the Fifth American Peptide Symposium' p. 576. (M. Goodman and J. Meienhofer, eds.), John Wiley & Sons, New York, 1977.
6. M.A. Ondetti, D.W. Cushman, E.F. Sabo and H.S. Cheung, *In* 'Drug Action and Design: Mechanism-Based Enzyme Inhibitors' p. 271. (T.I. Kalman, ed.), Elsevier, New York, 1979.
7. D.W. Cushman, H.S. Cheung, E.F. Sabo and M.A. Ondetti, *Biochemistry 16,* 5484 (1977).
8. D.W. Cushman, H.S. Cheung, E.F. Sabo and M.A. Ondetti, *Prog. Cardiovasc. Dis. 21,* 176 (1978).
9. D.W. Cushman, H.S. Cheung, E.F. Sabo and M.A. Ondetti, *Fed. Proc. 38,* 2778 (1979).
10. D.W. Cushman and H.S. Cheung, *Biochem. Pharmacol. 20,* 1637 (1971).
11. D.W. Cushman and M.A. Ondetti, *Biochem. Pharmacol. 29,* 1871 (1980).
12. D.W. Cushman, H.S. Cheung, E.F. Sabo and M.A. Ondetti, *In* 'Angiotensin Converting Enzyme Inhibitors. Mechanisms of Action and Clinical Implications' p. 3 (Z.P. Horovitz, ed.), Urban & Schwarzenberg, Baltimore-München, 1980.
13. M.E. Condon, E.W. Petrillo, D.E. Ryono, J.A. Reid, R. Neubeck, M. Puar, J.E. Heikes, E.F. Sabo, K.A. Losee, D.W. Cushman and M.A. Ondetti, *J. Med. Chem. 25,* 250 (1982).
14. I.A. Sytinsky, A.T. Soldatenkov and A. Lajtha, *Prog. Neurobiol.* (Oxford) 10, 89 (1978).
15. S.J. Enna and A. Maggi, *Life Sci. 24,* 1727 (1979).
16. G.A.R. Johnston, *Ann. Rev. Pharmacol. 18,* 269 (1978).
17. P.R. Andrews and G.A.R. Johnston, *Biochem. Pharmacol. 28,* 2697 (1979).
18. B.W. Metcalf, *Biochem. Pharmacol. 28,* 1705 (1979).
19. A. Nistry and A. Constanti, *Prog. Neurobiol. (Oxford) 13,* 117 (1979).
20. H. Möhler and T. Okada, *Nature (London) 267,* 65 (1977).
21. H. Möhler and T. Okada, *Mol. Pharmacol. 14,* 256 (1978).
22. S.J. Enna, J.F. Collins and S.H. Snyder, *Brain Res. 124,* 185 (1977).
23. N.G. Bowery, J.F. Collins, G. Cryer, T.D. Inch and N.J. McLaughlin, *In* 'GABA-Biochemistry and CNS Functions' p. 339 (P. Mandel and F.V. DeFeudis, eds.), Plenum Press, New York, 1979.
24. J.L. Waddington and A.L. Cross, *Neurosci. Lett. 14,* 123 (1979).
25. N.G. Bowery, D.R. Hill, A.L. Hudson, A. Doble, D.N. Middlemiss, J. Shaw and M. Turnbull, *Nature (London) 283,* 92 (1980).
26. J.S. Horng and D.T. Wong, *J. Neurochem. 32,* 1379 (1979).
27. S.J. Enna and S.H. Snyder, *Mol. Pharmacol. 13,* 442 (1977).
28. G.A.R. Johnston, R.D. Allan, P.R. Andrews, S.M.E. Kennedy and B. Twitchin, *In* 'Adv. Pharmacol. Chemother' Vol. 2, p. 11 (P. Simon. ed.), Pergamon Press, Oxford, 1978.
29. D.T. Wong and J.S. Horng, *Life Sci. 20,* 445 (1977).

30. P. Krogsgaard-Larsen, H. Hjeds, D.R. Curtis, D. Lodge and G.A.R. Johnston, *Neurochem. 32,* 1717 (1979).
31. P. Krogsgaard-Larsen, T. Honoré and K. Thyssen, *In* 'GABA-Neurotransmitters', p. 201 (P. Krogsgaard-Larsen, J. Scheel-Krüger and H. Kofod, eds), Munksgaard, Copenhagen, 1979.
32. A. Schousboe, *In* 'GABA-Neurotransmitters', p. 263 (P. Krogsgaard-Larsen, J. Scheel-Krüger and H. Kofod, eds.), Munksgaard, Copenhagen, 1979.
33. P.A. van Zwieten, J.C.A. van Meel and P.B.M.W.M. Timmermans, *Pharmaceutisch Weekblad (Scientific Ed.) 3,* 237 (1981).
34. A. Fleckenstein, *Ann. Rev. Pharmacol. Toxicol. 17,* 149 (1977).
35. P.D. Henry, *Am. J. Cardiol. 46,* 1047 (1980).
36. C.R. Triggle, V.C. Swamy and D.J. Triggle, *Can. J. Physiol. Pharmacol. 57,* 804 (1979).
37. K. Jim, A. Harris, L.B. Rosenberger and D.J. Triggle. *Europ. J. Pharmacol. 76,* 67 (1981).
38. J.C.A. van Meel, J.Q. Qian, P.B.M.W.M. Timmermans and P.A. van Zwieten, *J. Pharm. Pharmacol.* in press.
39. J.C.A. van Meel, A. de Jonge, H.O. Kalkman, B. Wilffert, P.B.M.W.M. Timmermans and P.A. van Zwieten, *Europ. J. Pharmacol. 69,* 205 (1981).
40. J.C.A. van Meel, A. de Jonge, H.O. Kalkman, B. Wilffert, P.B.M.W.M. Timmermans and P.A. van Zwieten, *Naunyn-Schmiedeb. Arch. Pharmacol. 316,* 288 (1981).
41. P.A. van Zwieten, J.C.A. van Meel, A. de Jonge, B. Wilffert and P.B.M.W.M. Timmermans, *J. Cardiovasc. Pharmacol. 4* (Suppl. 1), S19 (1982).
42. P.B.M.W.M. Timmermans, A. de Jonge, J.C.A. van Meel, M.J. Mathy and P.A. van Zwieten, *J. Cardiovasc. Pharmacol. 5,* 1 (1983).
43. R. Towart, E. Wehinger, H. Meyer and S. Kazda, *Arzneim.-Forsch. (Drug Res.) 32,* 338 (1982).
44. K. Satoh, T. Yanagisawa and N. Taira, *J. Cardiovasc. Pharmacol. 2,* 309 (1980).
45. P. Bellemann, D. Ferry, F. Lübbecke and H. Glossmann, *Arzneim.-Forsch. (Drug Res.) 31,* 2064 (1981).
46. P. Bellemann, D. Ferry, F. Lübbecke and H. Glossmann, *Arzneim.-Forsch. (Drug Res.) 32,* 361 (1982).
47. K.M.M. Murphy and S.H. Snyder, *Europ. J. Pharmacol. 77,* 201 (1982).
48. P.A.J. Janssen, C.J.E. Niemegeers, K.H.L. Scellekens and F.M. Lenaerts, *Arzneim.-Forsch. (Drug Res.) 21,* 1234 (1971).
49. P.A.J. Janssen, C.J.E. Niemegeers and R.P.H. Marsboom, *Arch. int. Pharmacodyn. 214,* 92 (1975).
50. A. Doenicke, E. Wagner and K.H. Beetz, *Anaesthesist 22,* 353 (1973).
51. A. Doenicke, J. Kugler, G. Peuzel, M. Laub, L. Kalmar, I. Killian and H. Bezecny, *Anaesthesist 22,* 357 (1973).
52. J.J.P. Heykants, W.E.G. Meuldermans, L.J.M. Michiels, P.J. Lewi and P.A.J. Janssen, *Arch. int. Pharmacodyn. 216,* 113 (1975).
53. G. Wahlstroem, H. Buech and W. Buzello, *Acta Pharmacol. Toxicol. 28,* 493 (1970).
54. B.T. Emmerson, L. Thompson and K. Mitchell, *In* 'Purine Metabolism in Man II – Physiology, Pharmacology and Clinical Aspects' p. 320 (M.M. Muller, E. Kaiser, J.E. Seegmiller, eds.), Plenum Press, New York, 1977.
55. J.D. Irvin, P.H. Vlasses, P.B. Huber, J.A. Feinberg, R.K. Ferguson, J.J. Schrogie and R.O. Davies, *Clin. Pharmacol. Ther. 27,* 260 (1980).
56. B.A. Brooks, E.M. Blair, R. Finch and A.F. Lant, *Br. J. Clin. Pharmacol. 10,* 249 (1980).
57. C.E. Wilhelmsson, J.A. Vedin, C. Moerlin, P. Lund-Johansen, C. Vorburger, W. Enenkel, P.M. Lutterbeck, J. Bolognese, V.J. Cirillo and K.F. Tempero, *Br. J. Clin. Pharmacol. 8,* 261 (1979).
58. J.A. Tobert, V.J. Cirillo, G. Hitzenberger, I. James, J. Pryor, T. Cook, A. Buntinx, I.B. Holmes and P.M. Lutterbeck, *Clin. Pharmacol. Ther. 29,* 344 (1981).
59. A.D. Ainley and H. King, *Proc. Roy. Soc., Ser. B, 125,* 60 (1938).
60. G.A.H. Buttle, T.A. Henry and J.W. Trewan, *Biochem. J. 28,* 426 (1934).
61. H. King and T.S. Works, *J. Chem. Soc.,* 1307 (1940).
62. C.C. Cheng, *J. Pharm. Sci. 60,* 1596 (1971).
63. P.-L. Chien and C.C. Cheng, *J. Med. Chem. 19,* 170 (1976).

64. M. Pailer and K. Porschinski, *Monatsh. Chem. 80,* 94 (1949).
65. A.R. Battersby and H.T. Openshaw, *J. Chem. Soc.,* 3207 (1949).
66. A.R. Battersby, R. Binks and G.C. Davidson, *J. Chem. Soc.,* 2704 (1959).
67. A.R. Battersby and S. Garratt, *J. Chem. Soc.,* 3512 (1959).
68. A. Brossi, A. Cohen, J.M. Osbond, P. Plattner, O. Schnider and J.C. Wickens *J. Chem. Soc.,* 3630 (1959).
69. A. Brossi, Z. Brener, J. Pellegrino, H. Stohler and J.R. Frey, *Experientia 16,* 62 (1960).
70. H.T. Openshaw and N. Whittaker, *J. Chem. Soc.,* 1461 (1963).
71. M. Barash, J.M. Osbond and J.C. Wickens, *J. Chem. Soc.,* 3530 (1959).
72. D.E. Clark, R.F.K. Meredith, A.C. Ritchie and T. Walker, *J. Chem. Soc.,* 2490 (1962).
73. M. Barash and J.M. Osbond, *J. Chem. Soc.,* 2157 (1959).
74. H.T. Openshaw, N.C. Robson and N. Whittaker, *J. Chem. Soc. C, 101* (1969).
75. A.P. Grollman, *Proc. Nat. Acad. Sci., US, 56,* 1867 (1966).
76. A.P. Grollman, *J. Biol. Chem. 243,* 4089 (1968).
77. N. Entner and A.P. Grollman, J. *Protozool. 20,* 160 (1973).
78. A.P. Grollman, *Science, 157,* 84 (1967).
79. L. Finch, *Europ. J. Pharmacol. 33,* 409 (1975).
80. P.A. van Zwieten, *Arch. int. Pharmacodyn. 215,* 104 (1975).
81. D. Wellens, J.M. van Nueten and P.A.J. Janssen, *Arch. int. Pharmacodyn. 213,* 334 (1975).
82. D. Wellens, A. de Wilde, A. van Bogaert, P.P. van Bogaert, L. Wouters, R.S. Reneman and P.A.J. Janssen, *Arch. int. Pharmacodyn. 215,* 91 (1975).
83. D. Wellens, L. Snoeckx, R. de Reese, R. Kruger, A. van de Water, L. Wouters and R.S. Reneman, *Arch. int. Pharmacodyn. 215,* 119 (1975).
84. D.G. Taylor and M.J. Antonaccio, *Clin. Exp. Hypertension 1*(1), 103 (1978).
85. P.B.M.W.M. Timmermans, F.P. Slothorst-Grisdijk, H.Y. Kwa and P.A. van Zwieten, *Arch. int. Pharmacodyn. 255,* 309 (1982).
86. P.B.M.W.M. Timmermans, F.P. Slothorst-Grisdijk, J.E. van Kemenade, A.M.C. Schoop, H.D. Batink and P.A. van Zwieten, *Arch. int. Pharmacodyn. 255,* 321 (1982).
87. P.B.M.W.M. Timmermans and P.A. van Zwieten, *Europ. J. Pharmacol. 61,* 385 (1980).
88. P.J. Beckett and L. Finch, *Br. J. Pharmacol. 74,* 190P (1981).
89. P.E. Hicks and C. Waldron, *Br. J. Pharmacol. 74,* 844P (1981).
90. P.B.M.W.M. Timmermans, Y.M. Harms, H.D. Batink and P.A. van Zwieten, *Biochem. Pharmacol., 31,* 3933 (1982).
91. P.B.M.W.M. Timmermans, Y.M. Harms, H.D. Batink and P.A. van Zwieten, *Neuropharmacology, 21,* 1039 (1982).
92. R.N. Brogden, R.C. Heel, T.M. Speight and G.S. Avery, *Drugs 16,* 273 (1978).
93. P.A. Baumann and L. Maitre, *Naunyn-Schmiedeb. Arch. Pharmacol. 300,* 31 (1977).
94. B. Harper and I.E. Hughes, *Br. J. Pharmacol. 67,* 511 (1979).
95. M. Raiteri, F. Angelini and A. Bertolin, *J. Pharm. Pharmacol. 28,* 483 (1976).
96. I. Goodlet, S.E. Mirlykes and M.F. Sugrue, *Br. J. Pharmacol. 61,* 307 (1977).
97. I. Cavero, R. Gomeni, F. Lefèvre-Borg and A.G. Roach, *Br. J. Pharmacol. 68,* 321 (1980).
98. J.R. Docherty and J.C. McGrath, *Naunyn-Schmiedeb. Arch. Pharmacol. 313,* 101 (1980).
99. B.B. Vargaftig, J.L. Coignet, C.J. de Vos, H. Grijsen and I.L. Bontje, *Europ. J. Pharmacol. 16,* 336 (1971).
100. H. van Riezen, R.M. Pinder, V.J. Nickolson, P. Hobbelen, I. Zayed and F. van der Veen, *In* 'Pharmacological and Biochemical Properties of Drug Substances' Vol. 3, p. 1 (M.E. Goldberg, ed.), Acad. Pharm. Sci. Pharmacol. Toxicol., New York, 1981.
101. J.H.R. Schoemaker, H.H.G. Berendsen, H.J.T. Stevens and V.J. Nickolson, *Proc. 21st Dutch Fed. Meeting,* Abstr. no. 380, 1980.
102. V.J. Nickolson and J.H. Wieringa, *J. Pharm. Pharmacol. 33,* 760 (1981).
103. V.J. Nickolson, J.H. Wieringa and A.M.L. van Delft, *Naunyn-Schmiedeb. Arch. Pharmacol. 319,* 48 (1982).
104. H.O. Kalkman, E.M. van Gelderen, P.B.M.W.M. Timmermans and P.A. van Zwieten, *Europ. J. Pharmacol.* in press.
105. H.O. Kalkman, E.M. van Gelderen, P.B.M.W.M. Timmermans and P.A. van Zwieten, *J. Pharm. Pharmacol.* in press.

106. H.O. Kalkman, Y.M. Harms, E.M. van Gelderen, H.D. Batink, P.B.M.W.M. Timmermans and P.A. van Zwieten, *Life Sci. 32,* 1499 (1983).

Racemates or Specific Isomers: Clinical Practice

K. H. Rahn

Abstract

There are now numerous drugs in clinical use of which the therapeutic effect is thought to be due to one specific isomer. This holds particularly true for beta adrenergic blocking agents all of which possess an asymmetric carbon atom in the side chain. Pharmacological studies have clearly shown that the beta adrenergic blocking action is almost exclusively due to the levorotatory isomer. Both, the levorotatory and the dextrorotatory isomer of some beta receptor blocking agents possess membrane stabilizing activity. Studies in patients with hypertension and in patients with coronary disease revealed that the beneficial effect of adrenergic beta receptor blocking drugs in these disease states is due to their beta adrenergic activity and not to the membrane stabilizing action. The same is true for side effects which can be observed when the usual therapeutic doses are given. Thus, in view of the therapeutic actions it would make sense to use only the levorotatory isomers of beta blockers. However, giving the dextrorotatory isomers concomitantly would not result in an increase of the frequency and the severity of side effects. On the basis of these considerations and in view of the sometimes cumbersome separation of the isomers, most beta receptor blocking agents are now used as racèmates for therapeutic indications. There are only a few exceptions such as penbutolol and timolol. The situation is similar for some other drugs.

There are now numerous drugs in clinical use of which a particular pharmacological activity is due to one specific isomer. Comparative studies with the different isomers make it possible to attribute a pharmacodynamic or therapeutic effect in man to a particular pharmacological activity. This can be illustrated by studies with beta adrenergic blocking agents all of which possess an asymmetric carbon atom in the side chain.

Propranolol, a beta blocker, has been frequently used by our group to study hemodynamic effects in normal volunteers and patients. Due to the asymmetric carbon atom in the side chain there exist a levorotatory and a dextrorotatory isomer of this drug. Numerous pharmacological studies have clearly shown that the beta adrenergic blocking action is almost exclusively due to the levorotatory isomer. However, both isomers possess membrane stabilizing activity. Animal experiments revealed that this may result in hemodynamic effects.

Propranolol as well as most other beta receptor blocking drugs decreases heart rate and cardiac output in man, both at rest and during physical activity. From a therapeutic point of view these hemodynamic actions are most important. Therefore, it seemed interesting to study whether they are due to beta adrenolytic or to membrane stabilizing activity. Previous comparative studies with the optical isomers of beta blockers had yielded no clear cut results. This could be due to the fact that fairly high doses of the drugs had been used and that a beta blocking activity of the dextrorotatory isomer given in high doses had not been definitively excluded. Therefore, we decided to reinvestigate this problem. For this purpose, only racemic and dextrorotatory propranolol were available in sufficient amounts. Our study was designed to compare the effects of racemic propranolol and of dextrorotatory propranolol on cardiac output, heart rate, venous and arterial pressures as well as on isoproterenol induced tachycardia[1].

The study was performed in volunteers without evidence of heart disease. Cardiac output was determined using the dye dilution technique. Heart rate was obtained from the continuously registered electrocardiogram. Arterial and central venous pressures were measured directly. The antagonism of the isoproterenol induced tachycardia was used to evaluate the degree of beta receptor blockade. Increasing doses of both the racemic and the dextrorotatory propranolol were injected intravenously. By this, cumulative dose-response curves for the above mentioned parameters were obtained. Previous studies had shown that in man intravenous doses of propranolol exert a constant beta adrenolytic effect 5 to 30 min after the injection. Therefore, all determinations were performed 5 to 30 min after the smallest dose of propranolol had been given.

The control value for cardiac output was 5800 ± 640 ml/min in the 6 volunteers who were given racemic propranolol and 6000 ± 620 ml/min in the 6 subjects given dextrorotatory propranolol. 0.5 mg Racemic propranolol lowered cardiac output by 9% ($p < 0.01$), 5 mg caused a decrease by 16% ($p < 0.025$). The highest dose of dextrorotatory propranolol decreased cardiac output, too, in the average by 15% ($p < 0.02$). The effect of 0.5 and of 1 mg of this isomer were not statistically significant.

Heart rate before the application of the drugs was 76 ± 3 beats/min in the subjects given racemic propranolol and 80 ± 6 beats/min in those given the dextrorotatory isomer. All doses of racemic propranolol significantly decreased heart rate, 0.5 mg by 7% ($p < 0.05$), 5 mg by 13% ($p < 0.005$). However, only the highest dose of 5 mg dextrorotatory propranolol decreased heart rate by 9% ($p < 0.01$), whereas the effects of 0.5 mg and of 1 mg were not statistically significant.

Neither racemic nor dextrorotatory propranolol changed stroke volume and arterial pressure. 5 mg Racemic propranolol slightly increased mean central venous pressure from 3 to 5 mm Hg. After 5 mg dextrorotatory propranolol this parameter rose from 3 to 4 mm Hg. Both effects were, however, not statistically significant.

Subsequently, two series of experiments were performed to measure the degree of beta receptor blockade. At first, in 6 volunteers the influence of racemic and of dextrorotatory propranolol on the tachycardia induced by 2 μg isoproterenol intravenously was determined. Racemic propranolol in doses from 0.5 to 5 mg clearly antagonized the isoproterenol induced rise of heart rate. Dextrorotatory propranolol slightly decreased the tachycardia by isoproterenol. However, this effect was not statistically significant.

In a second series, dose-response curves for the effect of isoproterenol on heart rate were obtained before and after application of a range of doses of racemic and of dextrorotatory propranolol to each of the 12 volunteers. Each day, only 1 dose of propranolol was tested in a subject.

Racemic propranolol in doses ranging from 0.125 to 5.0 mg intravenously shifted the dose-response curve of isoproterenol to the right in a parallel and dose-dependant way indicating beta receptor blockade. From the dose-response curves obtained, the amount of isoproterenol required to increase heart rate by 20 beats/min was determined. Under control conditions, no propranolol being given, 0.8 ± 0.1 μg isoproterenol had to be injected intravenously in order to increase heart rate by 20 beats/min. Significantly ($p < 0.05$) higher doses of the amine had to be injected to obtain this effect after racemic propranolol in doses from 0.125 to 5.0 mg had been given intravenously. In contrast, dextrorotatory propranolol in doses of 0.5

and 1.0 mg intravenously had no influence on the isoproterenol induced tachycardia. Only with the highest dose of 5.0 mg dextrorotatory propranolol had the isoproterenol dose to be increased to an average of 1.1 ± 0.2 μg ($p < 0.02$).

These two series of experiments demonstrate that dextrorotatory propranolol in a dose of 5.0 mg intravenously has beta receptor blocking activity. Furthermore, it appeared that studying dose-response curves for the isoproterenol induced tachycardia is a more sensitive procedure to demonstrate beta adrenolytic activity than to use a single standard dose of the amine. Quantitatively, the beta receptor blocking effect of 5.0 mg dextrorotatory propranolol was comparable with the action of 0.125 mg racemic propranolol. Thus, in these studies racemic propranolol is about 40 times as potent as dextrorotatory propranolol. Subsequent hemodynamic studies revealed that 0.125 mg racemic propranolol significantly lowered heart rate and cardiac output by 8% ($p < 0.05$).

We concluded from these results that propranolol decreases heart rate and cardiac output by means of a blockade of cardiac beta receptors and not through a direct cardiodepressive effect. The data suggest that the hemodynamic effects of 5 mg dextrorotatory propranolol intravenously also are due to beta receptor blockade.

Another problem that could be approached by studying the optical isomers of beta blockers was the question whether the antihypertensive action of these drugs is due to beta receptor blockade or due to membrane stabilizing activity. An early comparative study by Waal[2] with propranolol and quinidine suggested that the membrane stabilizing effect might be important. This was supported by a study of ours[3] where a single rather high dose of the optical isomers of N-isopropyl-para-nitrophenylethanolamine (INPEA) was given orally. However, a beta adrenolytic activity of the dextrorotatory isomer was not definitively excluded. In order to clarify the situation a study was started designed to compare the antihypertensive action of racemic and of dextrorotatory propranolol in hypertensive patients[4].

From an outpatient clinic, 10 patients with essential hypertension were selected to participate. All gave their informed consent. Based upon observations in the outpatient clinic, the blood pressure of these patients was assumed to be lowered by propranolol doses ranging from 60 to 480 mg daily. All diuretics and all antihypertensive agents except for propranolol were withheld starting at least 4 weeks prior to the study.

The study was of double-blind cross over design. Placebo, racemic propranolol and dextrorotatory propranolol were given orally for 4 weeks each in randomized order. The doses of racemic and of dextrorotatory propranolol given during the study were identical with the doses of racemic propranolol assumed to lower blood pressure during observation in the outpatient clinic prior to the study. This was in the average 240 mg daily. The total daily doses were equally divided into 3 single doses which were given at 8.00 a.m., 2.00 p.m. and 8.00 p.m. All measurements were performed 2 to 4 hours after a single dose.

During the study, the patients were seen at weekly intervals in the outpatient clinic. Arterial blood pressure and heart rate were determined twice at each visit after 10 min rest in supine position and after 2 minutes standing. The data obtained at the end of the first week of each experimental period did not differ from the values measured at the end of the fourth week. Therefore, the mean values from each period of a given patient were used for further analysis.

The number of tablets not consumed was counted each week in order to control patient compliance. Based upon this tablet count, the patients took 91 to 119% of their doses.

At the end of the third and fourth week of each experimental period, blood was drawn for fluorometric estimation of propranolol plasma levels immediately after the measurements of blood pressure and heart rate had been completed. Furthermore, the degree of beta receptor blockade was determined at the end of each experimental period using the antagonism of the isoproterenol induced tachycardia.

Racemic propranolol decreased supine systolic blood pressure from 170 ± 6 to 145 ± 4 mm Hg ($p < 0.001$). Diastolic blood pressure fell from 104 ± 3 to 94 ± 3 mm Hg ($p < 0.01$). In contrast, the dextrorotatory isomer did not lower supine blood pressure values ($p > 0.05$).

Racemic propranolol also lowered standing blood pressure data as well as supine and standing heart rate whereas the dextrorotatory isomer had no effect on these parameters.

It was considered that the differences in antihypertensive activity between racemic and dextrorotatory propranolol could be due to different plasma levels. However, during treatment with racemic substance, propranolol plasma levels averaged to 48 ± 6 ng/ml. The plasma concentration of propranolol was 49 ± 7 ng/ml at the time when the dextrorotatory isomer was given. The difference is not statistically significant ($p > 0.8$).

The intravenous dose of isoproterenol required to increase heart rate by 20 beats/min was determined from dose-response curves. During the placebo phase 0.8 ± 0.1 μg isoproterenol had this effect. The dose of the amine had to be considerably ($p < 0.005$) increased to an average of 27 ± 4.7 μg during the period when racemic propranolol was given, indicating beta receptor blockade. The data obtained during treatment with dextrorotatory propranolol did not differ from the values of the placebo period ($p > 0.05$). Thus, no beta adrenolytic activity of this isomer could be demonstrated.

No side effects were noticed during the period when dextrorotatory propranolol was given. In contrast, one patient complained of lightheadedness and another one of cold extremities during the phase when racemic propranolol was given.

This study demonstrated that in hypertensive patients whose blood pressure was decreased by racemic propranolol the dextrorotatory isomer had no antihypertensive effect. The differences in antihypertensive activity of racemic and of dextrorotatory propranolol could not be explained by different plasma levels. We concluded from the study that the antihypertensive effect of propranolol is due to beta receptor blockade. Meanwhile, this assumption is strongly supported by the fact that a number of beta receptor blocking agents with and without membrane stabilizing activity do exert a hypotensive effect.

Comparative studies with the optical isomers of beta blockers have also been performed to demonstrate that the renin lowering effect of these agents is due to beta adrenolytic activity and not due to a membrane stabilizing action[5,6].

As for hypertension, studies using the optical isomers of beta receptor blocking agents have shown that the beneficial effect in angina pectoris is due to the beta adrenolytic activity[7,8].

On the basis of the data available at present one may conclude that the therapeutic effects of these drugs in hypertension and in angina pectoris are due to their beta adrenergic blocking activity. Thus, it would make sense to use only the levorotatory isomers. On the other hand, there are no data supporting the assumption that giving the dextrorotatory isomers concomitantly increases the frequency and the severity of side effects. The limited experience from studies where these isomers had been used separately suggests that they even don't have any adverse effect at all. Based upon these considerations and in view of the sometimes cumbersome and expensive separation of the isomers, most beta receptor blocking agents are now used as racemates for therapeutic indication. There are only a few exceptions such as penbutolol and timolol.

Besides beta adrenergic blocking agents some other drugs are used therapeutically in form of their active optical isomers. This is true for a number of sympathomimetic agents such as noradrenaline. The amino acids dopa and methyldopa are given as L-isomers as only these are decarboxylated by the stereospecific dopa decarboxylase. Both, dopa and methyldopa have to be used in fairly large amounts in Parkinsonism and in hypertension, respectively, which sometimes results in local irritation of the gastrointestinal tract. Reducing the amount of substance given orally thus means less side effects. The thyroid hormones thyroxine and triiodothyronine also usually are given in form of their active L-isomers. Interestingly enough the D-isomers of these hormones exert a quantitatively similar effect on plasma cholesterol levels as the L-isomers, whereas their effect on heart rate and on oxygen consumption is considerably less pronounced. This led to the use of the D-isomers of the thyroid hormones as drugs for the treatment of hyperlipidemias.

References

1. F. Kersting, A. Hawlina, B. Lamberts und K.H. Rahn, Verh. dtsch. Ges. Inn. Med. *80,* 260 (1974).
2. H.J. Waal, Clin. Pharmacol. Ther. *7,* 588 (1966).
3. H.J. Gilfrich, K.H. Rahn and F.W. Schmahl, Pharmacol. Clin. *2,* 30 (1969).
4. K.H. Rahn, A. Hawlina, F. Kersting and G. Planz, Naunyn-Schmiedeberg's Arch. Pharmacol. *286,* 319 (1974).
5. J.A. Tobert, J.D. Slater, F. Fogelmann, S.L. Lightman, A.B. Kurtz and N.N. Payne, Europ. J. Clin. Invest. *2,* 309 (1972).
6. A. Philippi, U. Walther, A. Distler und H.P. Wolff, Verh. dtsch. Ges. Inn. Med. *80,* 263 (1974).
7. P. Björntop, Acta Med. Scand. *184,* 259 (1968).
8. A.G. Wilson, O.G. Brooke, J.J. Lloyd and B.F. Robinson, Brit. Med. J. 1969 II, 399.

Summing up

W. Soudijn

It is obvious that a living organism depends for its functioning and survival on regulatory mechanisms and biochemical processes that are stereoselective and even stereospecific in regard to the interaction of small endogenous ligands and biopolymers like enzymes and receptors. So it is not surprising that the stereochemistry of drugs and other bioactive compounds like e.g. pesticides, herbicides and poisons are of prime importance.

And although we all are aware of this - or at least ought to be - we are often only paying lip-service to this principle, as Ariëns in his well-known eloquent and inimitable way pointed out, when he said that we do not seem to mind spraying 500 kg of a racemic pesticide on say one acre of land when only one enantiomer is pesticidal while the other one is merely polluting the environment and that for economic reasons only.

Or when he and Ruffolo showed that we are in fact also polluting the internal environment (milieu interne) when we are clinically using an antihypertensive drug like labetalol with β- and α-adrenergic blocking properties, which has two centres of asymmetry and is a mixture of four compounds. The β-adrenergic blocking activity is caused by the compound with the R,R configuration while the α-adrenergic blocking property of the mixture is caused by the compound with the S,R configuration. The compounds with S,S or R,S configuration hardly contribute to the benificial effects and can be considered as pollutants.

From a practical and admittedly narrow point of view it is a pity that the α-adrenergic blocking activity is not caused by the compound with the S,S configuration as the S,S - R,R compound is often easily separable from the R,S - S,R compound, but for the understanding of the relationship between stereochemical structure and ligand - receptor interaction, labetalol or rather its R,R and S,R isomers can be very useful.

As an example of the dangers of pollution of the internal environment Ariëns mentioned the β-blocker practolol, a racemate, used in the treatment of hypertension but withdrawn from clinical practice because both the enantiomers of the racemate occasionally may cause severe and disturbing side effects. The useful β-blocking activity however, resides in one enantiomer. If this enantiomer had been used the toxic load would have been halved and thus the risk of obnoxious side effects would have been greatly diminished or even absent.

Vermeulen stresses in his paper the importance of the elucidation of stereoselective, be it substrate - product- or regioselective, biotransformation of drugs and other xenobiotics in relation to their pharmacological, therapeutic and toxicological actions in an early stage of drug development. In discussing stereoselectivity in mixed function oxidases or cytochrome P-450 systems he quoted the work of Jerina and co-workers who were able to design a computer simulated model valid in predicting the stereoselective metabolism of polycyclic aromatic hydrocarbons like benzo(a)pyrene, phenantrene, chrysene and benzanthracene, but had to conclude that such predictive models are as yet not available for other drugs.

Stereoselectivity in epoxidehydrolases and glutathione transferases seem somewhat more predictable, but in general predictability of stereoselectivity in drug metabolizing enzymes is rather poor. Species and gender differences in the rate of stereoselective biotransformations make extrapolation of data obtained from animal experiments to man risky as was demonstrated by the example of hexobarbital, the eutomer of which - S(+) hexobarbital - is metabolized more rapidly than the distomer R(–) hexobarbital in mice, rabbit, guinea-pig and rat, while in man the reverse holds.

The outlook seems brighter regarding the advances made in the analysis and separation of enantiomeric drugs and/or metabolites by HPLC and GC using commercially available chiral stationary phases or immunoassay methods for analysis in biological fluids, using enantioselective antisera.

Van Ginneken discussing stereoselectivity and drug distribution, states that, although carrier mediated transport mechanisms for endogenous compounds like glucose, ascorbic acid and amino acids are quite common, these mechanisms are rare for the transport of drugs.

Also drug-protein binding is generally not stereoselective for the enantiomers of a racemic drug. Apparent stereoselective drug distribution can mostly be explained as the result of selective biotransformation or selective excretion of one of the enantiomers, and not to differences in intrinsic distribution kinetics, so it is inferred from the data available up to now that stereoselective distribution does not play an important role in pharmacological of pharmacotherapeutic drug action.

Endogenous agonists interact with their receptors, recognition sites of (re)uptake mechanisms and active sites of their inactivating enzymes in different conformations. It may be useful to know these conformations in order to be able to design new drugs with a high selectivity for one of these sites of interaction. An approach of this problem is the use of rigid or semirigid conformers of endogenous agonists as Mutschler and Lambrecht showed on muscarinic agonists using cyclic acetylcholine analogues and their sulfoderivatives.

Together with the data presented by Dahlbom on the stereoselectivity and mode of binding of muscarines, muscarones, dioxolanes and their de-ether derivatives, an interesting succint overview of the relationship between stereochemical structural requirements and muscarinic cholinergic activity is obtained. In the second part of his contribution Dahlbom covered the requirements for stereoselectivity of tertiary muscarinic agonists and antagonists with a special reference to the fascinating, meticulous work on chiral oxotremorine analogues done in his laboratory, leading to the proposition that in the oxotremorine series agonists and antagonists probably bind to the same receptor site, in contrast to most antimuscarinic agents, which probably bind to allosteric sites on the muscarinic receptor. That NMR can be a powerful tool for the solution of stereochemical binding problems was shown by Burger when he described a.o. the elucidation of the mode of binding of the isomers of 4-amino benzoylglutamate and folinic acid to the enzyme dihydrofolate reductase and showed that part of the molecules binds in a similar fashion in both isomers and that the changes in the binding mode are caused by another part. Another interesting example is the elucidation of the methotrexate binding to dihydrofolate reductase by a combination of crystallography and NMR compared to the binding of the enzyme substrate folic acid.

Ruffolo presented an extensive overview on the stereoselectivity of adrenergic agonists and blocking agents and showed that the Easson-Stedman hypothesis (R(–) > S(+) = deoxy) holds for agonistic phenylethylamines with one centre of assymmetry interacting with all adrenergic subtypes, but does not hold for the imidazol(id)ines.

Phenylethylamines like e.g. α-methylnoradrenaline are active on all adrenergic subtypes in the 1R, 2S(–) erythro configuration. In disubstituted phenylethylamines like ephedrine the 1R, 2S(–) erythro form is preferred for storage vesicles and uptake I mechanisms.

Both α-adrenoceptors and uptake I recognition sites prefer the phenylethylamines in the extended trans conformation.

The stereochemical requirements for β-adrenoceptor antagonists are similar to those of the agonists.

From Bogaert's data on stereoselectivity in dopaminergic and antidopaminergic agents it appeared that although much work has been done on the stereochemical requirements in different dopamine receptor containing systems discrepancies between the results obtained in ligandbinding, adenylatecyclase assays and pharmacological tests do not yet allow an agreement on the classification of dopamine receptors and compounds that are acting more selectively like e.g. agonists that improve Parkinsonism without eliciting vomiting and nausea and antagonists useful in the treatment of schizophrenia without extrapyramidal side effects are badly needed for clinical reasons and for a better understanding of the molecular pharmacology of dopamine receptors. Witter's paper focussed on the changes in conformation and increases in metabolic stability and the effect on activity when D-aminoacids are inserted in peptides like LH-RH, the enkephalins and ACTH-neuropeptides. Although structure activity relationships are complex and a rational approach is not possible yet, sophisticated trial and error methods have been proven valuable.

Timmermans discussed the topic of stereoselectivity in various drug fields ranging from sexattractants to antidepressants. Attention was paid to the stereospecificity of angiotensin converting enzymes, gabaergic drugs, calciumentryblockers, antimalarials, emetine, hypotensive benzodioxane derivatives and mianserin and the importance of the stereochemical configuration for drug activity was clearly demonstrated. Although often a clear-cut explanation in molecular terms of the biochemical or pharmacological effects is at present not yet possible, it is evident that research in this field is challenging and will hopefully yield interesting results in the near future.

Rahn presented the results of his clinical pharmacological experiments with d.l-propranolol and dextropropranolol in volunteers and in patients with essential hypertension and concluded from these results and from data from the literature that the clinical efficacy of β-adrenergic blocking agents in hypertension and angina pectoric is due to the β-adrenergic blocking activity residing mainly in the levorotatory isomers and not to the membrane stabilizing properties of both isomers.

And this brings us to the problem whether to use eutomers or racemates in clinical practice. If the benefit of the presence of the distomer is nil, the risk of using the racemate is unpredictable because the distomer or its metabolites may interact unfavourably with a drug of comedication or have an unfavourable effect on a concomitant disease or provoke allergic reactions in some patients. This is not to say that racemates should never be used in clinical practice, as I hope to have demonstrated in my contribution.

My list of preference for the clinical use of drugs would be: 1. drugs without centre(s) of asymmetry, 2. eutomers instead of racemates, 3. racemates if superior to 1 or 2, for example, when both enantiomers have useful pharmacological properties in the right ratio, also because the registration board looks upon racemates as being one drug instead of a mixture of two different drugs, which in the latter case would make registration virtually impossible.

The development of new pharmacological tools – compounds with and without centre(s) of asymmetry, semirigid or rigid – will be necessary for an even better understanding of receptors, recognition sites of transport mechanisms and mechanisms of action, which in turn may subsequently lead to the design of better drugs.

A fertile field could possibly be found in the development of drugs with a partly peptide partly nonpeptide structure based on bio-active endogenous peptides as already shown by the design of angiotensin`converting enzyme inhibitors and stable enkephalins with semirigid ring closed structures.

Subject Index